The Story
of
Gardiner of Selkirk
1945 - 1988

"We're not selling cloth, we're selling fashion ingenuity"

Ewen McNairn

"Cloth is a by-product; what we are here to make is profit"

Colin Brown, to a meeting of the
Council of The National Association
of Scottish Woollen Manufacturers.

GARDINER'S TWEED ADVENTURES

Ian Jackson

Published by Ian Jackson, Farthing Green, Gattonside, Melrose TD6 9NP
Printed by Meigle Printers Ltd., Tweedbank Industrial Estate, Galashiels TD1 3RS

Ian Jackson 1998

First published in 1998 by
Ian Jackson
Farthing Green
Gattonside
Melrose TD6 9NP

ISBN 0 9534146 0 4

For Joan who, in 1954, left the Cheshire she loved for a place she had never heard of.

CONTENTS

Acknowledgements

Foreword

Acknowledgements

To everyone who has ever worked at or been connected with Gardiner of Selkirk Ltd. or its predecessor Edward Gardiner and Sons Ltd., for providing the story.

To the Directors of S. Jerome and Sons Ltd., for allowing me access to the minutes, records and files of correspondence which prodded my memory, and for allowing me to use Board papers as vital parts of the story.

To Margaret Duncan, of the Scottish Woollen Industry, for lots of help with dates and people.

To Geoffery Richardson and the National Wool Textile Export Corporation for access to the wartime reports and to details of the Korean war wool price trauma.

To the Scottish College of Textiles, and particularly to the library, for access to industrial records.

To numerous Gardiner employees in 1989/91 who filled the blanks in my memory.

To the numerous friends and competitors who, mentioned or not, put the butter on the bread.

To Colin Brown, for frequent responses, sometimes delayed, to the question "who was it that?".

To the late Edward Aglen for his thorough commentary on the first draft.

To Jim Whelans, for his comments on the draft.

Gardiner's Tweed Adventures

The Story of Gardiner of Selkirk Ltd.

Foreword

This is the story of one Tweed Mill through forty three years of dramatic change, not only to itself but to the industry of which it is a part.

In 1956 the National Association of Scottish Woollen Manufacturers, the trade association to which Gardiner belonged, had ninety one members, the majority being family businesses, with an office in Charlotte Square, Edinburgh.

In 1998 the Association no longer exists, being now part of an organisation embracing weaving, spinning and knitwear firms, with a tiny office in Glasgow. In July 1998 the disbandment of even that organisation was announced. Today only nineteen of the ninety one still exist even counting some swallowed up in greater empires and really unidentifiable.

The tale includes the mundane figures recording progress, and sometimes lack of it, but more importantly the human struggles of the management and workforce battling not just with rapidly changing markets but with each other, with customers and suppliers and most importantly with fellow shareholders.

Not all shareholders have the same objectives. The battle of the objectives includes chicanery, blackmail, starvation tactics and deviousness. The result of this unsavoury brew was, by 1988, a thoroughly prosperous and dynamic firm in a disappearing industry.

Life in the textile industry was never easy.

Chapter One

Selkirk

Climbing up the North West facing slopes of the Selkirk Hill and looking across the Ettrick Water towards Foulshiels Hill, Peat Law and Linglie Hill sits the small Border town of Selkirk. It is the second smallest of the Border towns with a population of some six thousand which makes it half the size of Galashiels, six miles to the North and a third the size of Hawick, twelve miles to the South. Peebles, at the edge of the Borders proper, away to the North has seven thousand souls but is now almost a suburb of Edinburgh. Only Langholm, away to the South towards Carlisle is smaller with a mere two and a half thousand folk: but what Langholm lacks in numbers it more than makes up for in entrepreneurial vigour. Which is not to say that Selkirk is exactly sleepy.

Between the Hill and the River lies a substantial tract of flat land, the Haugh. Between the junction of the Ettrick and Yarrow waters above the town lies Philiphaugh where Montrose, Charles I's Captain General, ran out of luck and in the fog and all kinds of alleged foul play saw his hitherto successful campaign come to bits about his ears, never to be recovered. On the flat at the foot of the Selkirk Hill the woollen mills and associated factories grew up in the nineteenth century, quite largely but not entirely as overspill from nearby Galashiels. The railway, which terminated at Selkirk and which had but one station on the branch from Galashiels ran along the Haugh a hundred yards or so from the river. Inland again from what was the mill lade, a canal whose waterflow was created by the cauld or weir, just below the Selkirk Bridge which carried the Peebles and Moffat roads, this lade provided power for the waterwheels of the mills until they were replaced first by steam engines, whose coal was brought by the railway, and, later, by electric power. The lade ran through or under most of the mills and with the advent of electric power it's principal use was as a source of poached salmon.

The Ettrick water flows northward and, after clearing the town, skirts the flat policies of Sunderland Hall at the northernmost point of which it joins that far more famous river, the Tweed, and loses it's identity. The merged, enlarged, river flows south of Galashiels, where it picks up the Gala Water, and on past Bemersyde and Dryburgh Abbey; past St. Boswells and Makerstoun and Floors Castle, seat of the Dukes of Roxburghe, where it is joined by the river Teviot, flowing down Teviotdale to Hawick and thence onwards to Kelso. Yet again enlarged the Tweed flows on under Rennie's handsome bridge and on by Coldstream to the sea at Berwick. Only Kelso and Coldstream of the Border towns were not substantially involved in the woollen trade from the nineteenth century on, both being centres of the agricultural industry.

Many of the Border towns have an annual festival, often known as the "Common Riding" and Selkirk's is the oldest. It celebrates, if that is the right word, the return of the thirteen "Souters" who survived at Flodden field. Held on the first Friday after the second Tuesday in June of each year it sees a massive cavalcade of horsemen, six hundred strong in recent years, ride out of the Town Square, down the Green and, fording the river at Linglie, riding out into the hills around the town's boundaries to return to the Square after fording the river again. In the Square the Standard Bearer "casts the colours", the cavalcade disperses and a substantial portion of it repairs to the various "Foy's" which take place in the public halls of the town. There is drinking, and singing, and the reciting of poetry, and storytelling for as long as chaps are prepared to go on, which is quite a long time! The run up to the great day includes a series of "ride outs" to meet the principals of the celebrations in the neighbouring towns, and a plethora of parties and balls. Souters regard the Common Riding as the peak of the social calendar, the prime event of the year. Some almost regard it as their raison d'etre. On a fine morning the cavalcade pouring out of the Square at half past six, and fording the Ettrick a few minutes later, is a magnificent sight. It looks a shade less sprightly on it's return after eleven miles or so of riding in the hills.

It was to this town that Edward Gardiner came in 1867 to set up in business as a woollen manufacturer. He had briefly been in business in Galashiels, and indeed kept his residence there at Langhaugh

House, but it was at his premises on Dunsdale Haugh, or Mill Street, known to this day as Tweed Mills that the business was really established. Gardiner was a weaver, having no spinning or dyeing machinery and remained thus throughout his life, by the end of which he had added another mill to his property, Linglie Mill at the northern end of the "industrial estate", again for the purpose of weaving only. The term weaving only is not exactly true, for the Gardiner mills carried out all the preliminary processes of winding, warping and drawing, and the subsequent mending and wet and dry finishing processes as well. They were weavers in the sense that, with no spinning capacity, they had to create their collections of fabrics from yarns bought in from others.

Gardiner was not the first on the Haugh at Selkirk. Roberts, at Forest Mill, were there long before him as indeed were others. Nor was he the last, but by the end of the century the mill buildings as they were in the nineteen thirties were in place. Many no longer house woollen mills. Some have been pulled down. Some have been put to other uses as diverse as a mushroom farm and a fish farm. Some are still at their original trade. When this tale of Edward Gardiner and Sons Ltd. is taken up in the nineteen thirties the businesses and their mills, from North to South along the Haugh, comprised the following.

Brown Allan & Co. Fine spinners in Riverside Mills.

Edward Gardiner & Sons Ltd., Weavers, in Linglie Mill.

Gibson & Lumgair Ltd., Vertical manufacturers in St. Mary's Mill.

Ettrick Mill. Weavers and Spinners

Edward Gardiner & Sons Ltd. Weavers in Tweed Mills.

Palfrey's Dyeworks, across the railway from Tweed Mills.

Craig Brown, in Yarrow Mill. Spinners.

George Roberts & Co. Ltd. Vertical manufacturers in Forest Mill.

The Heather Mills Co. Ltd. Vertical manufacturers, in Whinfield Mill.

Bridgehaugh Dyeworks, Dyers.

George Roberts & Co. Ltd. Spinning, in "the top mill", Philiphaugh.

The Corbie Lynn Co. Ltd. Weavers, across the river in Corbie Lynn Mill.

Somewhere among these, or in the Town were the premises of Baron and Hogarth, whose business was hiring "reeds", the grid or grating which keeps the warp threads in order, to the weaving mills. There were engineers, electricians and builders whose trade depended largely on the mills and all in all the population of Selkirk was almost wholly dependent on the woollen industry as were the outlying districts on agriculture. And, despite the fact that the origins of the industry in the Borders lay in the sheep farming of the district, by the turn of the century the wool used came almost entirely from abroad. From New Zealand, where there was a strong Border presence, from Australia, and the Cape, and from South America, with rarer fibres such as Mohair and Cashmere arriving from places as diverse as China, Turkey and Texas.

And what of the people of Selkirk, sometimes unkindly described as the only graveyard in the country with shops in it? People who, despite their great contribution to populating the Empire, particularly the Antipodes, though there is a Selkirk in Manitoba, have been known to say that a day away from Selkirk is a day wasted. With the whole of the town's fortune in the hands of the woollen industry they were, as one would expect, highly skilled in the crafts required by it, providing a marvellous fund of expertise for the firms involved to draw upon. Hard working and hard playing. The pubs in the town do not lack custom. The rugby team has it's ups and downs in the annual struggle with the other Border Clubs but is, none the less, a central feature of the town's sporting life with places in the first fifteen much sought after. The Golf club plays over a nine hole course on the Hill, notable for the fact that there are so few holes where you can see both the green and the intervening territory from the tee. But it breeds good golfers. W. Dickson Smith was, whilst Managing Director of Heather Mills and a member of the Selkirk club, fifth in the 1957 Open Championship at St. Andrews.

With great rivers close at hand fishing is a favourite sport and shooting in the moors and hills a popular pursuit. The cricket club plays on one of the pleasantest grounds in Britain at the site of Montrose's defeat and each summer organises "The Factory Cup", played for on a knockout basis by teams from the mills and trades. Originally an eleven a side, 25 overs each way affair, it is now down to seven a side, as is the Factory Cup for Rugby, played for each spring. The flying club keeps pigeons, not aeroplanes and the Opera puts on a show of high quality every year. Almost every taste is catered for in some way, and most at every age level. Television has dented, but not destroyed, the sporting nature of the people.

Hawick, twelve miles away, is a foreign place and the natural enemy. Galashiels, six miles away is if not an enemy then not an ally. Town loyalty is strong.

This is Selkirk, home of Edward Gardiner and Sons Ltd., whose tale I take up in the middle of the nineteen thirties.

Chapter Two

Between the Wars

The years following the first world war did not deal kindly with the Tweed Industry of the Borders, and Edward Gardiner and sons Ltd. was no exception to the general experience. By 1932 the Gardiner family were effectively out of the Company which was under the control of and in the main owned by Colonel Malcolm Murray Thorburn.

Edward Gardiner, the second, died on 30 June 1931 and as the last Gardiner to be employed by the Company, having served it as Managing Director for many years. Though the Company had prospered in better times, allowing the Gardiner family to join in the great building programme of Manufacturer's mansions, their contribution being Langhaugh House in Galashiels, times were now hard. But not so hard as to deny the late Edward Gardiner's widow a pension of £25 per month, not ungenerous in those days but, if the Board minute is accurate, promoted by the Company's self interest, "In view of the work the late Managing Director performed for the company and the harm which would result should his widow's circumstances be reduced, it was unanimously agreed in virtue of the power conferred by Section 17 of the objects clause of the memorandum to give the widow of the late Edward Gardiner, secundo, an allowance of £25 per month, in advance, from 1 July 1931.

Board minutes in early 1932 were devoted mainly to recording share transfers largely flowing from Edward Gardiner's death, but the accounts of the year to 31 January 1932 were approved at a meeting on 23 May of that year, and a meeting a week later records that they showed an accumulated deficiency on the Profit and Loss Account of £18,113, 18s, 7d of which £10, 352, 0s, 9˚d flowed from the year under review. A sad state of affairs for a company founded in 1867 but reflected in many a balance sheet in the industry. On the death of the late MD the balance sheet had been "restructured" and, despite the deficiency on the Profit & Loss account was noted as being "much improved". Perhaps creative accounting predated the nineteen sixties!

It was to be 1941 before the Company returned to profit. A smallish profit of £1963, 16s, 4d, which reduced the accumulated deficiency to £42,867, 11s 5d, a poor, if not alarming position, but none the less one from which Gardiners can be said to have, but for minor hiccoughs, gone forward consistently. The minutes of the intervening years are not very informative and tell little of the struggles to keep the company afloat. They do not record that Thorburn was an innovator and a man of vision, being one of the first to put substantial effort into the women's wear trade, the whole industry being, at that time, men's wear oriented. More than that he was early into co-ordinated cloths designed to create a ensemble of coat, suit and skirt in cloths of different weights. He had a long wait for his reward, such as it was.

The thirties were fraught with problems, not least in the important American market. All sorts of innovative ideas were tried there and the efforts to promote business incorporated features which, in the wrong hands, imposed unreasonable costs and risks on the mills. One such venture by a number of mills working through one American Distributor, Amalgamated Textiles, of New York, involved giving sample lengths free and loading the cost onto bulk deliveries, were they forthcoming. Whatever else was involved it all ended in tears, with a scheme of arrangement between nine British manufacturers (four of whom are still in business) and Amalgamated Textiles, providing for deferred payments of the amounts outstanding with a view to keeping Amalgamated afloat. Though the detailed agreement, showing the Gardiner share to be £2,900 of a total of some £80,000, is included in the minutes there is no record of the ultimate outcome. Did Amalgamated stay afloat? Did the mills get paid? In any event by 1946 Gardiner had nothing to do with Amalgamated.

Reporting on the accounts for 1932 the Directors "regret to report that trading has again been unsatisfactory and that the year closes with a loss of £4,702, 18s 8˚d", which is of course added to the debit balance. The minute of 28 September 1933 hints of domestic strife in the Gardiner family, for the widow's pension arrangement is reviewed and the original minute is supplemented "to the effect that the allowance was intended to be and is now recorded that as from 1 April 1933 it is for the benefit of the children as well as for Mrs. Gardiner to be applied by her for her and their maintenance".

The conditions in which trade was being carried on were indicated in the minutes of a meeting in November which sought the bank's consent to an increase in the overdraft limit from £32,000 to £35, 000 and the bank's response, set out in a letter dated 16 November 1933, is illuminating.

" Dear Mr. Thorburn

Following our meeting yesterday, I today submitted to the Board your application to overdraw the Working Account temporarily to a maximum figure of £35,000. I explained to them the position in which you found yourself resulting from unfortunate experiences of the six months to 31 July and the even more unfortunate shrinkage of output in the following three months but that you are now much busier and have work on hand which will keep the looms occupied for the next eight or nine weeks, business which you are confident will result in a profit. The main need for additional facilities arises from the urgency of making a payment to spinners, particularly Messrs. Laidlaw and Fairgrieve, and also for meeting wages for the next few weeks.

The Board, whilst reluctant to see the accommodation even temporarily increased, were not unmindful of the efforts you have been making against adverse circumstances to put the business on a better footing and eventually saw their way to sanction the additional facilities, which we trust you will be able to repay by the month of March. It is the understanding, as arranged with you yesterday, that you will endeavour to get the Yorkshire spinners to accept a three months bill in respect of the indebtedness to them, and if you are successful in this the accommodation for the bank should be correspondingly less.

I trust that the business you have at present will yield a satisfactory profit and assist in eliminating the loss for the year".

This letter was signed by the General Manager of the National Bank of Scotland Ltd. and introduces the parallel problems of cash and yarn which will play a prominent part in this history.

The pious hope of profit was unrealised. The year to 31 January 1934 yielded a loss of £7099, 10s, 9d, to be added to the deficiency. At this time Henry Watson Towns, the Company's accountant, resigned his Directorship for no disclosed reason, staying with the Company and Walter Sinclair, who ultimately became Gardiner's London agent for Menswear fabrics was elected to the Board, which then comprised him and Colonel Thorburn.

The loss for the year to January 1935 was just over three thousand pounds and that for the following year £1590. Things were looking up! 1937's accounts showed another loss, £1256, and 1937 yet another £721. In that year the Board pre-empted future government actions and decided to retire everyone at 65 with a proviso that there be recognition of long service.. The minutes do not disclose the nature of this provision, but when I joined the Company in 1954 it was a pension of sixpence per week (old, new equivalent 2.5p) for each year of service.

Something went wrong in the year to January 1939 and Gardiner's sustained a loss of £5040, a regrettable reversal of an improving trend, and this was followed in 1940 by another loss of £2760. By that time Henry Towns was back on the Board, and it had been strengthened by the addition of one William Emond, the mill manager. The profit in 1941 was followed by another, £2644, the following year. A major event at that time was the settlement of a debt due to Harry J. Gardiner £1677, 14, 5, for the sum of £625 in full and final settlement".

The war, of course, changed everything about the wool textile trade and not the least of the changes was the "Concentration of Industry" scheme whose objective was to concentrate the production of the various mills into the buildings of and onto the plant of a lesser number. In effect it meant that some firms shut their premises and had their product made on commission by the firms which stayed open. It was not quite a commission production situation, because the producer was designated in the scheme. As will be seen later Heather Mills was concentrated into Gardiner for the duration of the scheme, and untangling it after the war was a bit of a problem.

1943's result was a profit of £938, and 1944's another of £901 and the closing stages of the war parallelled the closing stages of the inter war conduct of the affairs of Gardiner. The first of the significant changes, (apart from concentration) was the sale of Linglie Mill to Scottish Prepared Papers Ltd. for eleven thousand pounds in September 1944, a transaction which would improve the cash position markedly and bring capacity more into line with probable demand. Malcolm Thorburn continued to gather into his own hands the various small shareholdings which were scattered about the Gardiner family, none of whom had any active interest in the business.

Came the end of the war and, on 12 June 1945 the Board resolved to recognise and reward those who had left the Company to serve in the forces. After discussion the following minute emerged: -

" It was resolved and agreed that recognition of this service should be made and in virture of the power conferred by Section 17 of the Objects clause of the Memorandum this should be in the form of an issue of War Savings certificates in the name of each employee for a value based on 5/- for women per week from 1 September 1943 to date of release plus the necessary additional weeks to complete a certificate but not beyond the 31 January 1946."

At that time Colonel Thorburn was released by the bank from a three thousand pound guarantee which he had given "as credit accommodation is no longer required". A weight off his mind.

The stage was set for the sale of the Company and the retiral of M. M. Thorburn. The transaction was largely the brainchild of the next Managing Director, Ewen McNairn, who was already in the trade in Corbie Lynn mill. He persuaded the company which was then Scotland's biggest dyer, The United Turkey Red Company, of Alexandria in Dumbartonshire, and Selfridges Ltd, who need no introduction, which two companies had at the time a common Chairman, to buy Gardiner. The public justification of the deal was that Gardiner would make cloth, U.T.R. would dye it and Selfridge would sell it, with profit accruing to all. Like all such neat and tidy concepts the reality was different. No such transactions ever occurred.

By evening of 28 September 1945 Edward Gardiner and Sons Ltd. was owned in equal parts by Selfridges and U.T.R., Colonel Thorburn picked up a thousand pounds for loss of office and two thousand a year, one year, consultancy contract and the Board consisted of: -
George Archibald Laing, for Selfridges
James Clark Campbell, for U.T.R.
Robert McEwen Sharp McNairn, Managing Director.
With H. W. Towns as Company Secretary.

If the Gardiner reign was phase one, and the Thorburn reign phase two, the scene was set for the eleven years of phase three.

Chapter Three

Times begin to Change
1945/46

At the time of the change of ownership and management the physical aspect of the Gardiner mill was much as it had been for the last thirty years or more. The buildings stood on the flat of Dunsdale Haugh sandwiched between the mill lade to the south, across which access was provided by a level bridge carried on steel girders, and the railway line to the north. The eastern wall of the weaving shed was right on the boundary of a field owned by Ettrick Mill and the western boundary was marked by a dry stone dyke along the boundary with the Yarrow Mill property. A neat little package, well hemmed in and with limited room for expansion.

Beyond the railway line were the buildings of Palfrey's dyeworks, now defunct and serving as a store for the British Cotton and Wool Dyers Association Ltd. to support the operations of their two finishing mills in Hawick. Beyond that an empty field, a handsome row of trees, the riverside road and then the river beyond which lay Linglie and many another hill.

The buildings were almost entirely single story, the only exception being a two storey block at the front whose ground floor was occupied by parts of the finishing processes and whose upper floor changed use, if not with the seasons, then at intervals of less than ten years. The main block of buildings including the two storey aberration was of traditional whinstone construction with saw toothed roofs lit along one face, slated on the other and separated by troublesome valley gutters. The main block was L shaped, with the engine and boiler houses in the crook of the L, but turned into a sideways lying U by the existence of a range of first war vintage wooden buildings housing preparatory machinery.

Separate from this, across the mill yard and to the west of the main building was a handsome sandstone block containing, at the front, the management offices and showroom and, behind those, the sample room and warehouse for the packing and storage of finished cloth.

The whole range of buildings was gas lit and heated only by steam pipes or, in the office block, by good old cast iron radiators, the steam for the whole coming from the single, coal fired, Lancashire boiler, which did double duty for it also provided the steam to power the whole of the machinery.

 At Tweed Mills the basic plant was a shed of sixty mixed Dobcross and Hattersley looms of various ages, but none in the first flush of youth. These were preceded by equally venerable winding, twisting and warping machinery and followed by scouring, milling, cropping, drying and pressing machinery similarly past it's best. The whole of this plant was driven by belts and shafting all running back to the single cylinder steam engine which powered the whole mill. Two processes which before too long were to be mechanised to some extent, Drawing and Mending, were at that time entirely manual and both were highly skilled crafts.

The workforce worked a five and a half day, 45 hour week, with two weeks holiday and the statutory public holidays, which in those days did not include Christmas Day. I well remember my own first few years at Gardiner, when the idea of going to work on that day and hurrying home for a quick Christmas lunch before going back to work again was not looked on benignly.

By the standards of today, after forty four years of progress, this sounds like the description of a thoroughly run down mill, but it is not. Perhaps not up with the best practice of the wool textile industry Tweed Mills was more typical than exceptional of it's kind.

With the sale of Linglie Mill this plant had, at the close of the war, to support the production of Gardiner and of Heather Mills, until the latter could be sent back to their own Whinfield Mill. The achievement of that move was a matter of priority, not only for Gardiner but for Heather too for both companies were inhibited by the shortage of capacity until the move could be made.

That, then, was the backcloth against which the new management was to perform and an indication of the problems to be addressed without delay. By 28 November 1945 McNairn had set out what would now be called his "Business Plan" and on that date he laid it before the Board, the only intervening meeting being in October with it's only business being the capital reconstruction and, guess what, the cash flow forecast.

With Heather Mills still on site only 32 of the 60 looms were available to Gardiner and as McNairn was in the van of the move to lighter fabrics which has gone on from then to now, and is still going on, even sixty looms would be inadequate, for lighter fabrics require finer yarns and more "picks" or weft threads for each inch and, therefore, loom speeds being to all intents and purposes fixed, production is slower. The new fabrics were of an existing construction of long standing whose basic feature was the mixing of yarns twisted in opposing directions and known generically as crepes. They were in the main self coloured and largely piece dyed, and the novel feature of the Gardiner "Crepe Tweed" was that they were designed as checks and stripes as if they were the traditional product styled for the season and time at which they were aimed. They were not easy to make, for the construction of the cloth depended wholly on the proper mixing of right and left handed yarns which posed problems for everyone from spinner through yarnstore and warping and weft winding to weaving, not to mention the special care required in finishing. This product was the "add on" which, allied to the traditional tweeds, was to take Gardiner out of the general run of Tweed manufacturers and provide the foundation for expansion.

The years of hard times between the wars had imbued the management with a false sense of the availability of labour and there is an assumption written into the minutes of that 28 November meeting that when Heather withdrew the labour would be available to go immediately to sixty loom output.

New machinery, particularly looms, was high on the agenda. The Crepes could best be made on Hattersley Light Standard Looms, and twenty of these were ordered, half each for 1946 and 47, and eight Standard Looms for the same split delivery. A Hattersley High Speed Warp Mill was also ordered for 1946 delivery.

But none of this was deemed quick enough and a proposal to buy 12 looms from the Botany Mill Weaving Company in Galashiels, for re-sale to them not less than one year later, was also approved. The total costs of the new machinery were £8510 and the net costs, after re-sale and removals, of the short term machinery were estimated at £1500.

I do not intend to spatter this book with figures such as these which, unless the reader is prepared to track back through the indices and revalue them at the date of reading, are meaningless. But it must be said that against the background of the Company's finances at the time these were substantial figures.

Electrification, both of lighting and power, was discussed. There was no dispute as to the desirability of both, but financial constraints had to be recognised. The new cloths required marked additions to the yarn stocks and expansion would generate increases in work in progress and debtors, but McNairn used the new, difficult to make, cloths as the basis for a compelling argument for installing electric light throughout the mill at a cost of £3300. But he did not win, getting Board consent only to improve the lighting in the weaving shed immediately, effective for this season.

The arguments for the application of electric power to the winding and twisting departments were: -
1.   To provide flexibility in these difficult to balance departments, and -
2.   To take the excessive load off the steam engine and boiler, and -
3.   To eliminate the considerable power transmission loss.

These arguments were accepted and the installation of three motors and equipment for these purposes was approved.

Advertising was the final plank in the plan as developed so far, with the re-establishment of the Company's name and products in overseas markets, neglected during the war, a priority. The Board agreed to twelve full page, full colour advertisements in International Textiles, the then "in" textile magazine, the advertisements to appear in alternate months commencing March 1946. The cost of this, not far short of £2000, was an earnest of the Board's determination to put Edward Gardiner and Sons Ltd. firmly back on the road.

The proceedings of this meeting can be said to be the first, tentative, step in a programme of re-equipment which has gone on uninterrupted to this day, and of a similar programme of expansion which has proceeded with only one significant setback from that day to this.

At a time like this, when wholesale change was the order of the day, it was quite natural for the Board to hold frequent meetings. If not monthly, then nearly so. And they soon discovered that if wishes were horses, beggars would ride, for on 10 January McNairn reported a labour shortage "from which all Scotch Tweed Mills were suffering", and went on to report on the impossibility of importing labour for lack of accommodation. It was agreed that he should discuss with the proprietor of an unnamed local hotel the prospect of acquiring the hotel for use as a hostel, and also reported that he was examining the possibility of renting premises in Eyemouth, forty miles away, with a view to setting up a mending plant there.

In other respects the Board's plans seemed to be going forward well. The first report that the overdraft would exceed the agreed limit was received calmly, the lighting improvements were under way and the orders for new machinery had been placed. Machinery was being moved to make room for the twelve looms coming from Botany Mill and the basis of the deal had been changed to a cheaper hire and return arrangement. Apart from the labour problem the only blot on the landscape was a difficulty in getting the three electric motors, and the initial power electrification would therefore have to start with temporary motors. The plan was nevertheless moving along as expected.

Fourteen sales agents had been appointed, all working on a commission basis and in countries as far ranging as Argentina and India. Not one of these is still with the Company. Death took some of the individuals, but successive tranches of sackings for failure to perform took the rest in the nineteen fifties. Of all the ambitious list of countries only Canada and the USA remain markets of any consequence at all, an interesting commentary on the changing patterns of trade in wool textiles.

There were two interesting hangovers from the past discussed at this meeting. Firstly it was agreed that the Company's holding of 3% Defence Bonds be sold, to help reduce the overdraft and the matter of small holdings of Preference Shares in Holland and Sherry Ltd. and A. Gagniere Ltd. was discussed. These holdings arose from the hard times of the past, when both these firms, London Woollen Merchants of high repute, had persuaded their suppliers to swap debt for Preference as a means of conserving their cost. Now that Gardiner needed the cash (it always had needed the cash) McNairn was authorised to try to negotiate the redemption of these Preference shares.

Though the minute refers to the Managing Director's report on sales and orders no detail is provided, but the Company was undoubtedly going forward at this time. More space is given to the cash position and the reasons for it's deterioration, though at this time the position was by no means serious and gave no cause for alarm.

As the year moved into spring the ideas came thick and fast. A Church Hall at Eyemouth was found and leased and staffed with local labour to mend fifty pieces per week and thus began Gardiner's long connection with that town. The Botany Mill looms were received and installed. A proposal to hire a bus to collect workers from outlying districts, workers who could not otherwise take employment in the mills, and return them home each evening was approved and with modifications, ran for twenty years. Another bright idea was to lease the Town Hall at Eyemouth, sound insulate it from the bank next door, and instal 12 looms to run double shifts. The arithmetic looked good, but it hung on the Sheriff agreeing to an amendment of the Feuing Charter for change of use. A three year lease was proposed, and the attempt to lure the Manager of the Commercial Bank in Eyemouth out of the Bank to manage Gardiner's Eyemouth operations failed.

"Unforeseen difficulties" prevented the Town Hall scheme getting beyond the bright idea stage, but it was replaced by another, to buy a site from the Eyemouth Town Council and build a building on it.

This one got through the Dean of Guild Court, got a building licence, was priced at £2804 and, after problems with the supply of bricks and pleas to the Ministry of Labour for help, drifted on till 18 September before being abandoned in the face of all the difficulties posed by the post war environment.

In June the Board approved the accounts for the year to 31 January 1946 which showed a profit of £1513.

Chapter Four

Post War Constraints

The post war trading climate was hedged about by constraints and controls stemming from the disruption of trading patterns brought about by the shifting demands and restrictions caused by the hostilities.

These had grown up over the years since 1939, had changed as the war's demands had changed and were still in force in the early post war period.

The four main constraints are foreign to today's market driven economy and the quartet comprised Price Control, Yarn Rationing, Export Allocations and clothing coupons, or clothing rationing. They took some years to dismantle and references to them crop up in the minute books of the time. As do references to another regime which restricted freedom of action, the balance between Utility Quotas, ordinary "free" Home Trade and Export Allocations. Later generations have found it hard enough to cope with ordinary market forces without these obstacles, but against that it must be recognised that the early post war period was a genuine sellers market, which made the constraints manageable.

In the early days of the war, when the weapons and supplies required had either to be made at home or bought from abroad and paid for in cash an export drive by those industries whose product was not in too heavy demand by the war effort was essential. The pre-war procurement of cloth for use by the Armed Forces had been efficiently conducted so Wool Textiles were free to play a prominent part in that drive.

To that end Sir Cecil Weir, K.B.E., MC. was deputed to organise the export effort of the Wool Textile Industry and he organised the meeting which led to the formation, on 6 March 1940, of "The Export

Group for the Wool Textile Industry", a body which not only fulfilled it's original purpose but which still functions today.

The early objective was to ensure that export effort was directed to those countries where Sterling would be required to pay for war materials. This basis of operation held good until 11 March 1941 when the United States Congress passed the Lease Lend Act which removed the need to provide immediate cash to the major supplier of weaponry and material. But Lease Lend did not have the unqualified support of the American People and it was not long before their industries, including their Wool Textile Industry generated a not unfounded complaint that Britain was using Lease Lend to bolster export trade to the detriment of the United States. Assurances were given that this would not be so and in the case of our Wool Textile Industry this assurance was implemented by the transfer of a significant part of the workforce to the arms producing industries.

The fall of France made the task of allocating export effort yet more difficult, and the entry of Japan into the war on 8 December 1941 so changed the picture that the export drive had to be run down to free manpower and shipping for more direct forms of war service. The activities of the Export Group were financed by a levy of one tenth of one percent on purchases of wool, a levy which was to persist for many a year as the objectives of the Group changed with the times.

The contraction of the capacity to export created it's own seller's market, particularly in countries where the perception of a lengthy cutback in supply prevailed. Whilst responsibility for determining the total quantity available for export rested with the Board of Trade, the Export Group were consulted in the allocation of the available goods to the most appropriate markets, and became entirely responsible for devising an equitable scheme for allocating those quotas among interested manufacturers. In the early days there was a lifeboat of supplementary licences for firms whose allocations did not cover existing orders, and for the export of goods already made for markets now closed, such as China, these latter being known as "frustrated exports". The Annual Report of the Export Group for 1942/43 summarises, in one of it's paragraphs, the difficulties of the times: -

" During the whole of the time under review, workers were continually being withdrawn from the industry to meet the requirements of the Armed Forces and the programme of munition and aircraft production. Though the priority claims of the Armed Forces and the coupon controlled home trade were rigorously scrutinised and pruned, the gap between demand and the reduced supply of which the progressively shrinking industry was capable, had to be filled by reducing the claims of export trade. In June 1943, after issuing allocations mostly on a reduced scale for Empire markets, and markets like the Middle East and for belligerent allies dependent on the United Kingdom for supplies, the Board of Trade indicated that there would be no allocations for the United States of America or any Latin American countries. Though this decision was hotly contested by the Executive Committee of the Export Group, it was upheld by the Government on the grounds that no production could be spared unless it was essential for the prosecution of the war by the United Nations".

By the time that report was written the war was going in the Allies favour and the closing paragraph foreshadows the immediate and future need for relief supplies of cloth to the liberated countries. And it includes reference to another little upset with our transatlantic friends. After America's entry into the war they began to fear that wool supplies would be inadequate to clothe their seven million man army and their civilian population and proposals were put forward for the compulsory blending of non wool fibres. The American manufacturers took the view that if they could not make pure wool fabrics, then their import should be prohibited. In the event the blending idea was never put into practice, and the squall blew over. But this incident does serve to illustrate the protectionism which has dogged exports to America over the years.

The 1944/45 report refers indirectly to the controls which existed on raw materials in alluding to "more export of British wool to North America", and to the start of release of wool to France and Belgium, "but not yet through normal channels". The continuance of Export Allocations is the subject of a paragraph which sets out very clearly the modus operandi: -

" Exports of wool piecegoods have followed the well known pattern laid down for them by the allocation system. The Board of Trade, after ascertaining from a War Cabinet subcommittee how much cloth can be spared for export purposes, fixes the quantities which can be allocated for each market. From this point an organisation controlled solely by the trade, namely the Export Group's Export Licensing Advisory Section, takes over and divides the quantities available for each market amongst exporters who have previously traded with that market".

The 1945/46 report covered the period during which both the war with Germany and that with Japan ended, and refers to the unfortunate need to continue with export allocations.

" Continuance of restrictions is the inevitable consequence of inability to supply all demands, especially in the case of an industry such as ours which supplies basic human needs rather than luxuries. In other industries not only have export licensing restrictions been abolished but Cabinet Ministers have threatened dire consequences if export trade were not substantially increased, but these industries either produce luxuries which are not essential to basic human needs or have ample capacity to supply essential home requirements and to develop an expanding export trade. Not so with our industry where home trade demand has to be tempered by coupon rationing and the like before even the present modest exports can be assured.

Yet we may be thankful for small mercies; exporters are being allowed more freedom in choice of market. Wherever possible the Board of Trade are agreeing to the issue of allocations which leave the exporter free to decide in which market he will sell his tops or his yarn or his cloth. If the choice is wisely taken, exporters will be able to use their small allocations, like seed corn, in those markets where they hope to reap a worthwhile harvest in the years to come".

By this time the sale of wool by auction had been resumed and the report comments on the fact that whilst price control prevailed for the home trade, 86% of the industry's product, severe control in the case of utility cloths, no control of raw material prices was in force or proposed, merely a statement that given the amount of wool in hand and under Government control the powers available to the Joint Organisation should ensure that the price did not rise above the general level which the Government in their wisdom consider appropriate. Only export prices were uncontrolled, leaving firms with the difficult choice in a seller's market, between going for a quick killing or building goodwill. With only 14% of the product "free", this did not represent much freedom!

The Export Group continued to be responsible for export allocations.

By the time the 1947/48 report was written the picture had changed beyond recognition. The assumption that wool prices would not rise significantly had proved over optimistic and topmakers were getting favourable treatment in the distribution of export licences "to compensate for losses in the home trade due to rapid advances in the cost of wool and the inadequate home trade ceiling prices". Yarn licences were no longer by market and in the case of cloth the emphasis had shifted from fulfilling home trade needs to export to hard currency countries. Comparison of the industry's weaving capacity with the supply of clothing coupons led the Board of Trade and the Export Group to the conclusion that home supplies would not suffer if exports were virtually freed, and the allocation system at this time recognised this, with firms almost able to write their own licences. But yarn rationing was still a severe constraint, particularly in respect to worsted yarns. But all in all freedom was beginning to return to the industry.

By the following year the industry was smarting under criticism for failing to meet the Government's export targets, and attributing the failure to the inability of negotiators to overcome the problem of import quotas in customer countries and, ironically, the commodity with the best prospect of meeting those targets was yarn, where export licences were still restricted to ensure a supply of yarn to the weaving industry! Freedom had returned to that industry to the extent that export licences had been replaced by monthly declarations of quantities exported and analysis of these returns showed that the woollen merchant, traditionally a massive exporter of home produced cloth, was losing ground to direct exports by manufacturers and suffering from a stock build up. This naturally generated a spate of applications to sell "frustrated export" fabric on the home market, with the Board of Trade reluctant to encourage this trend.

By the sixteenth of March 1949 the only raw materials still subject to export licensing were rags, and these were cleared on 18 July. However, in order to stop the dollar drain resulting from the export of raw wool to the USA via the continent, limited export licensing was re-imposed on 19 September. All in all the problems had moved away from capacity and licensing to the tough nuts of price control and purchase tax as instanced by a paragraph in the Export Group's 1949/50 report: -

" The nation will expect our industry in the coming year to improve upon it's performance, both in production and in export, but maximum results will be obtained in the long run only if certain rigidities and distortions are removed. As we leave the war and it's aftermath behind, we see more clearly that a sound export trade needs a healthy home trade as it's basis; and yet home trade is violently divided. On the one hand the cloths, on which rests our reputation as producers of the best in the world, pay purchase tax in the home market at the rate of 66 2/3 percent. Saville Row and the haute couture in London cannot develop or even sustain their role as spearheads of world fashion unless they can be assured of a reasonable turnover. On the other hand, the utility programme does not fit into a situation in which steep increases in the price of raw material call for more flexibility in balancing home and export demands than arbitrary price ceilings will allow".

The reference to raw material prices covered an increase of twenty percent, a rate which was to pale into insignificance in the following year when the panic induced by the Korean War drove prices to a peak some 280% higher, only to fall, in the same year, to a mere 37% of the highest levels. It was providential that that trauma followed the freeing of the industry from most of it's restrictions, for it would have been far more disastrous had all the past controls been in force.

From then down to now the artificial constraints on the industry have been in the field of tariffs and quotas, the instruments which Governments still use to protect or hamper the free flow of trade, the choice of word depending on your point of view.

Chapter Five

The Early Post War Years 1946/48

In the next few years Gardiner progressed at a commendable pace despite a number of background factors which might have been specially designed to inhibit progress. The first and most important of them was the Company's under capitalisation, which probably underlay all the other adverse factors. At 31 January 1946 the total net worth was £34,704 and the net current assets £13,044, both figures brought to these low levels by the attrition of the thirties. From this lowly starting point the whole business of modernising the mill's machinery, it's buildings and it's product had to set off.

The modernisation of the plant alone involved a quantum leap from steam power and gas lighting to electric power and light incorporating all the things which had not been done in the thirties. The plant, venerable Dobcross and Hattersley looms and equally venerable pre and post weaving machinery had to be updated, largely replaced and made suitable for new products, particularly for the ambitious Crepe Tweed programme. The capital spend authorised over the calendar years 1946 and 1947 was as follows: -

20 Hattersley model 375 looms	£	5000
8 Automatic looms		2960
1 High Speed Warp Mill		550
Conversion to electric light		3600
Conversion from steam to electric power		3448
Office, lavatory, canteen and rest room		3000
Building at Eyemouth		2804
Austin Lorry		350
2 Twisting Frames		3300
House for Cost Accountant		2050
Company Car, Humber Snipe		1050
	£	28,112

Set against the background of a £19,000 deficit on the Profit and Loss account at the end of January 1946 and an historical depreciation charge of the order of £3000 a year, this is a very substantial programme, a fact which was recognised by the major shareholders who each agreed to put up a loan of £10,000 at 3.5% per annum. But mere survival demanded such investment, and progress would call for much more.

And though the major shareholders thus demonstrated their support the fact was that after the initial enthusiasm for the Gardiner project had worn off they were, not to put too fine a point in it, disinterested. Their representatives, very senior Directors in their own companies, attended Board meetings at approximately monthly intervals but neither company was familiar with the woollen industry. Selfridge is well known to most people, a large London department store which, in the post war era prospered. The United Turkey Red Company was a very large dyer and printer of cotton textiles operating in outmoded premises and with outmoded plant at Alexandria in the Vale of Leven. They, like Gardiner, were struggling for survival in the post war era and, unlike Gardiner, they ultimately lost their battle and for many years in the nineteen seventies and eighties their only asset was their Gardiner shareholding. It is not surprising, therefore, that they regarded Gardiner as peripheral and concentrated their efforts on keeping it under control, rather than pushing it into expansion.

Add to this mixture of a run down woollen mill and two major shareholders of very limited dedication the Joker in the Pack, the Managing Director, Robert McEwan Sharp McNairn, and the result is a potentially lethal brew of conflicting personalities and divergent aims and ambitions.

Ewen McNairn, as he was always known, was a designer at heart and by training, an acknowledged leader in his field at that time. Besides being a designer he understood the basic concepts of what is now known as "marketing" and was a consummate salesman who could charm the birds out of the trees. His years of experience running the Corbie Lynn Mill across the river in Selkirk had given him a sound background in manufacturing techniques and processes so he could be said to be a textile man through and through. If there were chinks in his armour they were twofold and inter-related. Firstly he was a very proud man who found it hard to acknowledge fault or error and secondly his grasp of finance was unsure. Perhaps it was not so much that it was unsure, but more that in the situation in which he found himself his natural instinct, given his pride, was to ignore those financial aspects of the business of running a woollen mill which did not fit in with his views.

At this time Ewen McNairn was unfortunate in that Henry Towns, the Company Secretary, was elderly, was unqualified, had grown up with the business and was no match for Ewen's personality, so rather than contributing to fill in the gaps in his dynamic MD's broad experience he lapsed into the role of bookkeeper leaving the finance function in a vacuum.

McNairn's design philosophy, which might well have been appropriate to a financially stronger company, did not include any element of product rationalisation. If he had an idea, and if a particular yarn fitted in with it, or if a yarn attracted him and could have a cloth built round it, the yarn was bought and this modus operandi inevitably resulted in a yarn stock which held, along with a limited number of basics moving at a more or less constant pace, a large number of "One Offs" which could move fast, slowly or not at all. One of his favourite dicta was "we're not selling cloth, we're selling fashion ingenuity", and the stockholding consequences of this outlook were to cloud the Company's financial position for many a year.

Not all the product was of this "Novelty" character. High hopes were pinned on the Crepe Tweed fabrics and the 20 Hattersley 375 Looms were bought specifically to make that product; indeed the minutes of the time record the fact that the older looms could not make it satisfactorily and this rigid plant configuration of twenty looms which could make Crepe Tweed well, but which could make little else in the product range, and the rest of the plant with exactly opposite characteristics was a recipe for future difficulty, for whatever the initial success of Crepe Tweed it has been the case, in the woollen trade, that nothing lasts forever. Not a recipe for future disaster, for alternative products could be and were developed to use the 375's when the Crepe Tweed ran out, but that particular block of looms did have a relatively limited useful life.

That cloth was nevertheless a very important factor in pushing forward the Gardiner fortunes at this time. It was novel and it was wanted and by October 1947 McNairn, full of enthusiasm for it, sought the Board's approval for the acquisition of additional machinery for it's production. "The Board considered the possibilities of limited and large scale expansion against the desirability of consolidation of technique and solid establishment in overseas markets at a time of world wide economic disturbance. The Board decided, despite the very favourable introduction of Crepe Tweed, to wait for a consolidation period before deciding on an expansion programme".

Given the financial constraints under which Gardiner laboured there was really no other decision which they could take, especially as the major shareholders were unwilling to fund such an expansion programme. Having thus demonstrated their financial orthodoxy the Board went on to show that they were, in other ways, chancers. Having bought a new car they decided to delay the sale of the old one "pending the Government's final decision on the basic petrol ration, which would considerably affect the sales value of a second-hand car".

The tightrope act of balancing the growing yarn stock against the overdraft, made more difficult by substandard production volumes as a result of labour shortages and the low production rate of the new Crepes continued to exercise McNairn and aggravate the Board. The task was made no easier by the need to specify in advance the proportions of Utility and free market cloths to be made in future periods, which reflected back on the yarn rations allocated to the Company. As these had to be specified in four monthly periods, four months ahead of first deliveries, it is easy to see how, should production lag, yarn stocks could run away should spinners keep to time. As they were not really expected to do so insurance at that end of the chain tended to be built in so a combination of suppliers performing and mill production underperforming, for whatever reason, could make the overdraft and underlying yarn stocks move very rapidly in the wrong direction. As early as May 1946 Gardiner decided to take only half of the usual Utility quota for the next rationing period in order to allow production and deliveries to catch up.

The backlog was not quickly cleared and in December 1946 McNairn, supported by the Board, decided to place yarn orders as usual for March/June 47 period, but to defer taking our own orders for the corresponding period till a later date. The Board's agreement to this, considering it's unavoidable impact on yarn stocks and the overdraft is, in retrospect, a little surprising. But at this early stage they were new to the game and had not yet worked out all the angles. And McNairn was a very persuasive man. Eight months later it all came home to roost.

The minutes of Board meetings had always been written by Ewen McNairn and not, as might have been expected, by the Company Secretary, and as his difficulties with production and yarn stock problems, and with the Board, increased, they became less and less a record of decisions taken and more statements of his opinions. The minutes of the August 1947 meeting exemplify this trend and, after attributing the December yarn orders/cloth selling decision the Board go on to stress that the MD had guessed right about certain changes in the yarn rationing arrangements, to the Company's benefit, and that the earlier decision had made it possible to take part in the Export Drive. But when the rhetoric is complete and the conclusions drawn it turns out that the overdraft is now £38,000 and will rise to £48,000 by mid September. Records do not show what the agreed facility was at the time, but the last recorded figure was £10,000 in October 1945. It is unlikely to have been more than doubled in the intervening period. The minute goes on to say how the introduction of plain Crepes has dropped output by 20% and that consumption of yarn has, of course, similarly fallen. The situation is succinctly set out, in contrast to the rest of the minute, in the sentence: - "As a result, yarn deliveries increased while cloth output went down".

Following that observation the record goes on to show that "the Board decided, therefore, at this meeting, that while the objective had been attained, the amount of yarn stock work in progress, and piece stock, were much too high, and that these should be reduced as quickly as possible, bringing down the Company's overdraft at the same time". Then follows the passage which will form the justification for the next set of difficulties, whatever form they may take. "The Board recognised that the Company would, throughout the next six months, be in a position to take the fullest advantage of the Export Licences available to it, but that further intake of yarn should be restricted while existing stocks were cleared". And the escape hatches are not yet all battened down for the minute also records that stock reductions should take account of the danger of standstill in the winter months "when suppliers might, through fuel shortage, have to cut down deliveries". As a throwaway at the end of the minute records the Board's view that by the end of January the overdraft should not exceed £25,000 to £30,000.

The yarn stock problem was to become a familiar one over the next ten years.

Prominent among the subjects debated was the question of representation, or how to sell the product. The industry had over the years used three main methods. Selling direct from the mill, the Directors being traditionally the salesmen to the mostly London Woollen Merchants, who then sold on in smaller quantities through their own sales forces. The employment of salesmen who sold only the mill's products to both merchants and garment makers and, lastly the use of Selling Agents who, for a commission, would sell the product in their specified markets. Such agents would represent the products of more than one mill, but provided there was no conflict were a good way of spreading the cost of having a good man on the ground. A Scotch Tweed manufacturer and a Yorkshire worsted manufacturer could share an agent without conflict, but the Scotsman could not share with a high quality Yorkshire, or West of England, Tweed maker. The cost savings did not come without problems attached.

Gardiner had for many years been represented in the London women's wear trade by agents, Alan Shepherd & Co. Ltd. who had offices in Great Portland Street in London. At that time the Directors of that agency were the ageing Alan Shepherd himself, and the young Ken Smith, fresh out of the RNVR soon to be joined by Gerald Box when he was released from the Indian Army. In January 1946 fourteen more agents were appointed in markets ranging from the USA (Folkard and Lawrence) to Argentina. Not one of these still represents the Company and few, perhaps only one, survive. Thus did Gardiner take up the agency route to sales. But left over from pre-war arrangements was part of Amalgamated Textiles, referred to in an earlier chapter.

This part was Cornwall Mills, a subsidiary of Amalgamated, registered at Tweed Mills, Selkirk. The AT operation had been as a "tied merchant" selling only the products of it's owners and Cornwall was perceived as being in the same field. McNairn visited the States in April 1945 and made contact with Cornwall, who seemed keen to develop a trade in the USA in fabrics originating in the UK. The Board appear to have taken the view that producing for and selling to Cornwall would not conflict with the activities of the newly appointed Folkard and Lawrence, but decided that current capacity did not permit of active participation, but that the next meeting should be provided with the records of Cornwall's past performance. At that subsequent meeting it was decided that co-operation with Cornwall should be offered starting in 9 to 12 months time, despite certain pricing problems which had been experienced in the past.

McNairn and Cornwall kept in touch and jointly developed Cornwall's plan to the point that a meeting between McNairn, John Lawrence of Cornwall, and Bernard B. Smith, Lawrence's lawyer, was convened in London to discuss that plan. It was that Cornwall should buy a mill of their own, "and it would appear from confidential information possessed by Mr. McNairn this might be achieved by Cornwall making an offer to the proprietors of a certain Scotch Tweed Mill". McNairn's report goes no to say that Cornwall appear to be enthusiastic and that Bernard B. Smith is making arrangements to make the offer. And also that he (McNairn) had indicated that in the event of the purchase going through we (Gardiner) could undertake the supervision of the mill at a suggested fee of £3000 per annum, payable in America, to which Cornwall had agreed.

But by 9 January 1947 the deal was coming to bits and McNairn reported to the Board "on a letter which he had received from Mr. Henry Booth, the text of which was most obscure and difficult to understand as a reply to this which contained a very full submission on the advantage of Cornwall acquiring a manufacturing unit of their own as had been recommended to them". The Board suggested that Mr. Booth be invited to come over to discuss the proposition and get an appreciation of the prevailing conditions.

By 12 March 1947 Mr. Blake Lawrence had, on a visit to Selkirk, advised McNairn that the project had been abandoned by Cornwall, and that is the last we hear of them.

In all the programmes of capital re-equipment it was noticeable that no provision was made for finishing machinery even though it was in no better state than the weaving, winding and twisting which was being replaced. That did not mean, however, that McNairn had overlooked the problem for in August 1946 he floated the idea that there was only one process, milling and scouring, that could not be done on the machinery at UTR, and that a solution might be to transfer milling and scouring machinery to Alexandria and have all the finishing done there. The Board thought highly of this suggestion (they would, wouldn't they; saves capital!) and called for cost comparisons.

By the next meeting in September UTR men had visited the mill to see what was involved and, in particular, had examined the milling machines to see what would be involved in their transfer. They obviously got at James Campbell, the UTR man on the Gardiner Board, for the plan was abandoned with only the proviso that trials be carried out at UTR to see how they fared. Later there is a reference to trials not yielding satisfactory results and the idea went away.

Reference has already been made to shortfalls in production. As much as anything these were due to labour shortages which were inevitable as the mills which had been closed under the "concentration of industry" scheme were re-opened. Different mills devised different solutions to this. Laidlaw and Fairgrieve built a mill in Dalkeith where there was a pool of labour. Ettrick Mill bused workers down from Dalkeith each day. The Gardiner solution was to send a bus up the Yarrow valley each day collecting workers from the outlying farms and villages and, later, to send another to Lilliesleaf, a village a few miles away in the other direction. During 1947 the bus was discontinued and an Austin lorry was bought to serve the same purpose and also to run pieces back and forth to the darning department at Eyemouth, forty miles away. The colonisation of the Yarrow valley, and of Lilliesleaf, were to last for many years and only succumbed to the combination of the reduced labour demand of more modern machinery and the near universal ownership of motor cars.

Other events in the period which takes us roughly to 31 January 1948 were the appointment of Mr. T. F. Donaldson as Cost Accountant at a salary of £450, and the provision of a house for him. The establishment of Directors fees at the rate of £300 a year with an extra hundred for the Chairman, a subject dear to James Campbell's heart. The concentration of advertising in the new monthly textile magazine "The Ambassador". Participation in the "Enterprise Scotland" venture, with special displays and cocktail parties at Jenners on Princes Street, Edinburgh. The decision not to insure home trade debtors and the resumption of dividend payments. And, last but not least, a Court approved Capital Reconstruction which wrote up the assets by £20,000, reduced the adverse balance on the Profit and Loss account by the same amount and left the Company with an issued capital of £50,000 with £20,000 of undenominated shares for issue at the Director's discretion. The ordinary shares were held as to UTR 22,500, Selfridges 22,500 and Ewen McNairn 5000.

The year to 31 January 1947 yielded a Gross Profit of £7,686 and the year to 31 January 1948 one of £19,754. An excellent result in those days, particularly against the background of the difficulties which had to be overcome and the limitations of margins on Utility fabrics. But it was still a seller's market.

Chapter Six

Relative Tranquillity 1948/1952

Gardiner's history is not over filled with periods of peace and quiet and the three years which followed the close of the year to 31 January 1948 form one of the longer spells of tranquillity, which is not the same as to say that they were without either incident or important developments.

Throughout the period the problem of representation in the United States market exercised McNairn and his Board. As early as July 1948 McNairn expressed dissatisfaction with Folkhard and Lawrence's performance and suggested the use of a French manufacturer who had established his own selling organisation in New York. The underlying problem, one which to this day has not really gone away, was a double headed monster. Firstly, Folkhard and Lawrence were primarily representatives of manufacturers of men's wear fabrics, and therefore skilled in their sale rather than the sale of the women's wear fabrics which were Gardiner's principal product. And, secondly, the United States market was much more receptive of men's fabric imports than of women's. Together these two factors were to force Gardiner to produce men's wear fabrics, largely abandoned in the post war era, in order to establish a presence in the lucrative American market. The French connection came to naught which was probably just as well for it might have delayed further the realisation of the need for men's wear fabrics.

By November 1948, with the Autumn 1949 Collection newly in New York the Board decided to let Folkhard's fate hang on the outcome of that season, a decision which held sway for a mere three weeks, for in December McNairn reported that Folkhard's reaction to the Collection was so negative that a change must be made. Since at this time McNairn could not be spared from the mill for a trip to the States to seek other arrangements or find out what was inhibiting the activities of the present agents it was agreed that the very experienced Alan Shepherd, the Company's London Agent, be asked to make the journey and find out why the collection had been so ill received. This visit took place in April 1949 and took on board both it's original purpose and that of giving assistance to Folkhard and Lawrence in the launch of the Spring 1951 Collection, one which had been confined to a limited number of clothes and styled in an entirely new way. The visit was reported on at meetings in May and July, but the minutes are brief and say no more than that the agency was to stay where it was. Another decision with a very short life.

At the Board meeting held the following month the minute under the heading "Representation in the USA", reads as follows: -

> " While it had been agreed at previous meetings that representation in the USA should be left with Folkhard and Lawrence Inc. there was doubt as to whether any great success in Women's Wear sales would be achieved through this".

McNairn went on to report informal discussions which he had had with Jacqmar Ltd. an important London customer who will appear more than once in this tale. They were in the process of opening their own merchanting operation in the States and the idea which had been bruited was that they should sell Gardiner's more expensive fabrics, as merchants, through the new organisation, and the less expensive ones as agents for a commission. Jacqmar had approved the idea in principle, subject to the agreement of the Gardiner Board and the start up of their new American operation. The Gardiner Board favoured the scheme and asked McNairn to keep in touch with Jacqmar. Thereafter this idea seems to have foundered without trace, for by 28 December 1949 McNairn reported that he had been forced to terminate the Folkhard agency. No reference to the Jacqmar proposal appears in those minutes but that meeting did generate the link between Gardiner and Lennox Knitwear Ltd.

This was a company set up by United Turkey Red to manufacture knitwear in Alexandria for sale, amongst other places, in the United States, for which purpose a New York subsidiary, Lennox of Loch Lomond, was set up. Since a selling organisation in New York set up to serve a brand new Scottish knitwear manufacturer was bound to be an expensive operation the idea of using the selling organisation to sell Gardiner's cloth as well had superficial attractions to UTR and to Gardiner, who were momentarily unrepresented in the USA. The idea does not seem to have been thought through and in the event the arrangement had a limited life. The principal objection to such an arrangement, particularly in the case of small companies such as Gardiner and Lennox is that the target customers are entirely different for the two products. Cloth needs to be sold to garment makers or cloth merchants, whilst knitwear is sold to retailers. To hope to find the connections and skills to the required degree in a very small sales force is asking too much. Nevertheless the idea went ahead and by March 1950 McNairn was able to report, on his return from a visit to the United States, that Lennox had acquired suitable premises, had furnished them and had appointed a Resident Director, one Drummond S. Musset, of whom more later.

Ever since the takeover in 1946 the Board had comprised Mr. G. A. Laing, Chairman, representing the Selfridge shareholding but wearing two hats, for he was also Chairman of UTR, Mr. James C. Campbell, representing the UTR shareholding and R. M. S. McNairn, the Managing Director. The Chairman was an Edinburgh Chartered Accountant, and James Campbell's family had their roots in the textile trade through the family business of Campbell's, Stewart's and McDonald's, a firm of Glasgow general textile merchants or "packmen", so called by reason of their travellers in former times going about the countryside carrying their wares in packs. He was a director of UTR, of the Standard Life Assurance Company Ltd. and of the family business which by then was much reduced from it's former glory. The Secretary was H. W. Towns, the same Towns who, before the takeover had, briefly, been a Director.

The first change in the composition of the Board came in May 1950 when the Chairman died, and when it was proposed that the late Mr. Laing's suggestion that Mr. Hugh Cowan Douglas, a Director of UTR be appointed to the Board to strengthen it "subject to the agreement of UTR and Selfridges". Not surprisingly, as such an appointment would have given UTR total control of the Board, Selfridge's agreement was not forthcoming, and the next meeting on 28 June sees the appointment of Mr. H. A. Holmes, the then Chairman of Selfridges Ltd. to the Gardiner Board with James Campbell moving into the Chair. This Board remained unchanged until 17 January 1951 when, as part of the arrangements for reconstructing the Company's capital, Hugh Watson, W.S., was appointed a Director with a view to his becoming Chairman, which he did on 5 March 1951. Hugh Watson was a highly regarded Edinburgh lawyer with many interests in industry who, in later years, became Deputy Keeper of the Signet. Though technically this office is that of active President of the senior society of Solicitors in Scotland, the Office of Keeper being an office held by a member of the Nobility, not by a solicitor, the Office of Deputy Keeper of the Signet is a very prestigious one. Inevitably a member of the New Club, Hugh Watson, was an active playing member of the Honourable Company of Edinburgh Golfers and was very active in the training and welfare of young people both of and outside his own profession. Hugh Cowan Douglas never did achieve a seat on the Gardiner Board.

The design team of those times was two strong. Alex Linton, approaching retirement, who never took his cap off in any company or circumstances, as head designer supported by Jack Dalgleish. The other member of the then management team was Robert Grey, the Mill Manager who, in early 1949 had to resign for ill health. In his place McNairn appointed a Yorkshireman, probably the best trained textile man of his time, Frank Schofield, who joined on 1 April 1949. The cost accountant, T. F. Donaldson, was still in that post.

These three years were a period of consolidation and prosperity. Pre tax profits were, for years to 31 January 1949 £20,975; for 1950 £21,817; and for 1951 £29,812; all of which were very good results for those times though in terms of today's inflated currency they are not obviously so. The tone of the period was perhaps set by the Board on 25 February 1948 when they laid down a policy of maximum liquidity consistent with the maintenance of turnover, restricted yarn stocks and work in progress and limited development in terms of machinery beyond that already planned. The results show that McNairn lived very successfully with these restrictions and they were slightly relaxed in February 1949 when more ambitious stocking of classic yarns was authorised. Machinery purchases in the period were minor; a reaching in machine for £350, Coning machines for £360, a van for £280 and a winding machine for £270. Lighting at the Church Hall at Eyemouth and more changes to the cloakrooms and toilets cost a little, but compared with the previous programmes this was just minor adjustment. Sundry disposals of worn out machinery, and of plant no longer required included unfortunately, the steam engine which had powered the mill before electrification. Unfortunate because had the space which it occupied not been urgently required for an engineer's shop it would have been a significant piece of industrial archaeology for future students of the trade. But the growth of output with corresponding growth of yarn stocks and work in progress put great pressure on space.

So much so that in April 1949 a plot of land at Eyemouth was bought for £100 to accommodate a proposed £3500 building. In the event this building was never put up and the plot was disposed of. Gardiner's Eyemouth operations remained in the Church Hall for many years and ownership of a building there had to wait until 1977. At the same meeting McNairn put forward the case for an extension to the existing Yarn Store building to house the twisting machinery due for delivery in April 1950. This first proposal was turned down by the Board who, "whilst not unmindful of the encouraging results achieved, were hesitant meantime about embarking on further capital outlay, having regard to the commitments yet to be taken up, and the problematical maintenance of turnover in markets the trend of which was developing in the buyer's favour". Even so they agreed that estimates for the extension be sought and in December the project, to cost £7000 was approved and by the following March it was building. The next "territorial demand" was not long in brewing for in the minutes of the 5 December 1950 Board meeting there is talk of a £4000 extension reportedly agreed at a meeting in August in connection with machinery deliveries. The minutes of the August meeting make no reference to either machinery or buildings, though they do record the major shareholder's agreement to put up temporary loans of £5000 each.

Profitable as it was Gardiner was still under capitalised, a condition evidenced by frequent references to control of the overdraft and intermittent temporary shareholder loans, and it was not until the end of this period of tranquillity and progress that the nettle was properly grasped, In late 1949 it was part grasped when the Authorised Capital was increased from 50,000 ordinary and 20,000 undenominated shares to £100,000 being 50,000 ordinary, 20,000 undenominated and 30,000 new 6% Preference Shares. On 8 November

1949 an Extraordinary General Meeting approved these proposals, converted the 20,000 undenominated shares into 6% Preference Shares and agreed to issue these 20,000 shares and 10,000 of the earlier Preference, 15,000 each to the two major shareholders. Following these transactions the issued capital of 50,000 Ordinary Shares and 30,000 6% Preference Shares was held as follows: -

Ordinary,	Selfridges	22,000
	UTR	22,000
	RMS McNairn	6,000
Preference,	Selfridges	15,000
	UTR	15,000

RMS McNairn had acquired 5000 shares in the 1948 reconstruction, and had bought 500 more from each of the major holders in the same year. The Selfridge and UTR shareholdings were, with true Victorian secrecy, held in the names of bank nominees.

But only part grasped was the nettle, for, with payments for buildings, machinery and the increased yarns stocks resulting from diversity flowing from the Company's increasing Novelty fabric trade the injection of £30,000, substantial though it seemed at the time, was but a finger in the dyke and a mere year was to elapse before the injection of capital which was to see Gardiner through, with frequent bumps in the road, to and beyond 1988.

As the year 1950 developed the need for further capital became more and more obvious as did, despite the progress the company was making, the reluctance of the major shareholders to commit further funds. Selfridges because they were disinterested, and didn't want to, and UTR because they couldn't. At this time the charge of disinterest cannot be laid at their door, for they were pretty active in their interest in the Company. There had been the abortive transfer of finishing idea, and the joint venture in New York was up and running, though it had not yet lifted off. William Calderhead, the Managing Director of UTR was taking a personal interest in both that venture and was doing what he could to further Gardiner's interests. A West of Scotland man, as were both James Campbell and Ewen McNairn, he was very well connected in the financial scene in Glasgow, and in particular with a company called Glasgow Industrial Finance Ltd. Today they would probably be described as a merchant bank, and it was to them that the Board went looking for £50,000 and it was they who arranged it's provision by fifteen Investment Trust Companies, all with good Scottish names, though seven of them had London addresses, and The Scottish Amicable Life Assurance Society. For introducing this capital GIF Ltd., charged a commission of threepence per share, £625 in total.

The resolutions and arrangements for reconstructing the Company's Capital were complex and were approved at a meeting held in the North British Station Hotel in Glasgow on 17 January 1951. Part of the deal was the conversion of the 30,000 of short lived 6% Preference Shares into Ordinary shares, making the Issued Ordinary Capital up to £80,000. The 20,000 unissued Preference Shares were converted into Undenominated Shares and 50,000 new 6% cumulative Redeemable Preference Shares were created. These were the shares to be issued to the Investment Trusts on 29 January 1951.

The terms of redemption were reminiscent of a lottery. Profits being available for the purpose 2000 shares were to be redeemed each 30 April from 1952 onwards, with an option for the Company to redeem more if it chose. The shares to be redeemed each year were to be chosen by drawing, which involved putting into a bag numbered tickets, one for each 100 shares lot, drawing twenty of them, identifying the shareholders concerned and then re-sealing the bag. It worked surprisingly well and there was only one hiccup when, after drawing 20 tickets, their listing only came to 1900 shares. In those days trousers had turnups, and James Campbell's turnups revealed the missing ticket, relieving the Board of the need to count the remaining tickets to see that 20 had actually been drawn.

If profits were not available for redemptions they accumulated, though the end date of 1977 for completion did not move. No redemption could be made if the Preference dividends were in arrear; that deficiency had to be made up first. In the course of time all these events occurred. Dividends and redemptions fell into arrear, were caught up and finally the balance of shares was redeemed ahead of time. The men who wrote the rules must have had a crystal ball.

The £5000 each of major shareholder's loans were repaid and at the end of the period of tranquillity the Issued Capital was:–

80,000 Ordinary shares of £1 each, held as to 37000 each by Selfridges and UTR and 6000 by Ewen McNairn.

50,000 6% Cumulative Redeemable Preference Shares, held by Investment Trusts and the Scottish Amicable.

Backed by 20,000 unissued Undenominated Shares of £1 each.

And, allowing for the redemption through time of the Preference Shares, so it remained until 1988 except for Bonus Issues which took the Issued Ordinary Capital up to £400,000 using the undenominated shares along the way.

The Capital reorganisation was accompanied by the drafting and issue of new Articles of Association which, like the Capital Structure, were to stand for many a year.

Though relatively tranquil these years were not easy, for the post-war seller's market ended and a more normal framework for the textile trade, the buyer's market in which it had mainly operated since Victorian times, returned. As early as July 1949 McNairn reported on the changes in trading conditions prevailing in all the Company's markets. But, as ever in a design and fashion oriented industry, changes in trading conditions do not apply universally across the board and a month later the minutes record the success of the latest Utility range, which had filled up the order book to and beyond the end of the year; a long order book indeed in the Women's trade.

Thanks to Gardiner's strength in design, and to the heavy investment in machinery in the post war years the firm was still prospering and going forward though the state of under capitalisation was making itself felt. At the December 1950 Board meeting McNairn had to report that weaving output was only running at 80% of capacity, and that this situation would prevail until the winding and twisting machinery currently being installed came on stream. The strength of demand for the Company's product is underlined by the concern expressed for the effect on Gardiner's reputation which late deliveries resulting from producing below target might have, and the upset which that might cause to continuity of production of the specialist fabrics on which success was based. The strength of the firm at this time is evidenced by the fact that calendar 1950 was to be followed by an even more successful calendar 1951 before the change in trading conditions hit home.

The Utility fabric arrangements restricted profit margins on those cloths and in April 1948 it was proposed that a subsidiary company be formed to distribute Edward Gardiner's products, which arrangement would have the effect of allowing an extra profit on fabrics "twice sold". This Company, Gardiner of Selkirk Ltd. was registered in September 1948 and thereafter all Edward Gardiner sales were to Gardiner of Selkirk who sold on to the ultimate customer. All profit figures quoted after that date are consolidated figures. Gardiner of Selkirk Ltd. had an Issued Capital of £25,000, all but one nominee share owned by Edward Gardiner and Sons Ltd. and so it remained until 1978 when the two companies swapped names and the subsidiary was wound up, it's usefulness having ended years earlier.

Naturally the new subsidiary took on the selling agencies and when the Lennox connection came to fruition took on that as well. The joint company ceased to be Lennox Knitwear and became Gardiner Lennox Scottish Woollens Inc. of New York, with Drummond Musset in charge at a salary of $10,000 a year and a commission of 4% on sales over $240,000. The salary, and the operating expenses of the New York Company were financed by a sliding scale of commissions paid by Gardiner; on the first $125,000 of sales of Gardiner cloth, 11% on the next $125,000, 5% and on sales in excess of $250,000, 4°%, with an underlying minimum commission, whatever the sales, of $12,500. The Gardiner minutes do not record what Musset's company was to get on sales of Lennox knitwear, and , with prescience, it was agreed that in the event of the termination of the agreement Gardiner and Lennox would each be responsible for half the liabilities. But for items dealing with the brief life of Gardiner Lennox the Gardiner of Selkirk minutes comprise only the formalities of Annual General Meetings until the Name changing and winding up.

Finally in this period, the Gagniere and Holland and Sherry shareholdings were disposed of, at substantial percentage but modest cash losses.

Chapter Seven

Crunch Time for Ewen McNairn
1952/53

The period reviewed in the last Chapter ended with Gardiner's year to 31 January 1952 which showed a record profit of £36,268 on turnover of £481,157, largely based on Home Trade sales. The American market being without satisfactory representation and other world markets failing to perform for the same reason. But things were already going wrong and decisions on products were about to be taken which by reason of their yarn stock implications and their limited suitability for the existing kit of machinery were to exacerbate the under-capitalisation and cause disastrous friction within the Board.

The word "recession" first appears in the minutes of the February 1952 meeting, and I quote below the bulk of a paragraph in which the Managing Director describes the immediate effects of the recession.

" . . . for notwithstanding the full Spring order book which kept production at peak during the Autumn months of last year, trade recession has also crept in and this was fully exploited by customers when delivery dates for goods came round. Customers endeavoured to get out of their commitments on the most trivial excuses, concessions had to be made, extended credit became rife and generally it was a most worrying time, as it still is. Apart from these finished goods troubles there were others to deal with in the form of yarn commitments and stocks, on which substantial writes down have had to be made to bring them down to the market level.

Notwithstanding the need to deal most drastically with all these provisions it is anticipated that the figure of profit should not be any less than last year's, at £35,000".
(In the event it was a little better than that at £36,268).

In common with almost the whole of the British Wool Textile Industry the United States market was ever in the minds of the management, and in Gardiner's case there was much to be done if a satisfactory result was to be achieved there. The problem of representation was, as usual, crucial and the Gardiner Lennox business was not living up to the hopes which had led to it's foundation a mere two years ago, so much so that in June of 1952 McNairn was reporting to the Board that representation in that country had deteriorated and that it was most unsatisfactory. That "Mr. Musset does not appear capable to meet selling problems and while every opportunity has been given him to develop ideas sponsored from this side the reaction to this has been nil". The Board considered ending Musset's agreement, but left the decision to follow McNairn's American visit in July and Musset's visit to Selkirk, in late August.

Clearly Drummond Musset was not wholly incompetent, and did have the power to formulate ideas and draw conclusions. Perhaps the seeds of the problems were in the difficulty of selling two disparate products to equally disparate markets, and of working for two masters. If the Lennox attitude to the development of ideas was, as Gardiner's appears to have been, limited to ideas originating from "this side" then he was likely to be, and remain, frustrated, for he was not a man to keep his opinions to himself. During his visit to Selkirk he was introduced to others in the trade and in particular spent a day at one of the spinners which provided yarn to Gardiner, Laidlaw and Fairgrieve in Galashiels. He was taken under the wing of Russell Fairgrieve, the scion of that family business. Having spent the day seeing the processes and problems of the spinning business, the whole of it in Russell's company, the pair returned to the Douglas Hotel for a well earned refreshment and in the course of that evening's conversation Musset said to Fairgrieve; "Well, young Russell, I know that besides being a director of Laidlaw & Fairgrieve you are very interested in politics. If my opinion will be of any help I suggest that you forget spinning and stick to politics". Forthright? Rude? Perhaps, but with a degree of prescience, for Russell Fairgrieve, having been instrumental in developing the Shetland knitwear trade which could be said to have been the saviour of the woollen spinning industry in the seventies and eighties, did almost give up spinning when Laidlaw & Fairgrieve was sold to the Dawson group, of which it remains a profitable member. He did go into Parliament as member for one of the Aberdeenshire constituencies, he did become Minister of Health for Scotland and bowed out of Parliament as Sir Russell Fairgrieve. Did anybody at Gardiner listen to what Musset had to say? Or were the underlying difficulties simply insurmountable?

By October the Gardiner Lennox saga had reached the point of no return, with Lennox giving notice of their intention to withdraw from the arrangement at a convenient date, leaving Gardiner to continue with a representation which had failed and in which McNairn had no confidence. There was nothing for it but to suggest to Lennox that a final settlement be come to, with the New York operation being closed down, Musset paid off and the assets liquidated, all of which was achieved with contention only over responsibility for payment of Drummond Musset's final commission entitlement. The furniture was sold, and the cloth in hand on consignment handed on to the new agent for disposal.

There were in New York at that time a pair of firms of Wool Textile selling agents run by one George Klein, one of the industry's dynamos. George Klein must have known of Gardiner's representational troubles (surely did, for the Wool Textile trade is a gigantic ladies tea party as far as gossip and rumour goes), and at the psychological moment he offered his services, or rather the services of one of his firms, Trans British Textiles Inc. to represent Gardiner in the United States and so began a longish relationship with a relatively short period of real success from Gardiner's point of view. A period in which an excellent working relationship developed between Gardiner and George Klein and his sidekick Henry Slome. A relationship which, by reason of the force of George's personality and his real knowledge of the American market, made Gardiner enter the US men's wear trade effectively for the first time since the war, and a period in which Trans British Textiles prospered to the point of being one of the leading agencies in New York, with three of the very top Yorkshire worsted manufacturers in their portfolio, Taylor and Lodge Ltd., Kaye and Stewart Ltd. and Broadhead and Graves.

But, as in so many ventures, setting the scene for future prosperity had it's short term costs and problems. Trans British Textiles were appointed Gardiner's agents in November following a meeting between George Klein and Philip Robinson in London at the beginning of that month. Philip Robinson was a new Gardiner director, appointed on 28 October 1952, nominated by Selfridges Ltd. to replace Mr. H. A. Holmes who had retired. Philip Robinson was to become a very active member of the Board. He had come up through the Lewis's stores business between the wars during that firm's period of dramatic expansion and had been involved in it's purchase of Selfridges Ltd. in 1952.

The costs and problems to be invested in future success made themselves plain at the first business meeting between Ewen McNairn and George Klein. George did what he had to do and specified the product which could both be made by Gardiner and sold by Trans British Textiles in New York, and it wasn't what Gardiner were currently making, being sports jacketing and suiting cloths made from expensive worsted yarns, and even more expensive cashmere yarns. It had to be done. McNairn recognised it. The Board recognised it, but it brought in it's train a whole new range of yarns to be stocked and financed and helped to bring right back into focus the under capitalisation of the Company and the limited suitability of the weaving plant for the principal trades in which Gardiner would be involved, fine crepe tweeds and relatively fine cashmeres and worsteds. Forty six of the looms were suitable, but 27 were not and the cost/production equation reared it's head with a vengeance. Making relatively fine cloths, 46 looms could not make a profit, but finances would not stand replacing the 27 obsolete machines let alone support the yarn stock and work in progress which they would generate.

Without doubt the appointment of a dynamic American agent, and taking his advice on product were fundamental to the survival and future prosperity of Gardiner, but the immediate consequences had not been thought out. Indeed the situation was typical of the textile trade, or probably any other trade. A run of success, with increasing profits, falters, in this case by reason of a slowing down of trade chiefly consequent on the Korean war, which had an unnerving effect on, amongst other things, wool prices. Overstocking of raw material resulting from the slow down and a change in the level of demand for the successful product resulting from imitation by competitors. In this case it was crepe tweed which suffered from imitation, and the efforts to keep it ahead of the competition by introducing more and more colours into the designs added to the stock problems. The other part of the women's wear product, novelty fabrics, continued successful, but by it's very nature it was capital intensive of yarn stock, depending as it did on novelty and variety. In short the product had become too diverse and the machinery of questionable suitability. Then along comes someone who convinces you that product X is the answer and even though it does not really suit your machinery or your method of cost allocation the offer is taken up, financial burdens are increased and the overall cost structure further distorted.

The Korean war, mentioned above, had unsettled the worlds wool textile industries. In those days the key indicator of wool prices was the '64s B top", theoretically this is wool, combed for worsted spinning, which can be spun to a count of 64s on the worsted system, that is to say 64 x 560 (35840 yards) to the pound. From mid 1948 to January 1950 the 64s top had risen, not evenly, from 119 pence per pound to 124. By May it was 171, by September 240, by January 1951 260 and it peaked in March at 348. The fall was equally swift, 260 in June, 195 in July, 132 in October before settling into a range of 152 to 119 for 1952. Everyone's perceptions of raw material costs and consequent selling prices were distorted and conditioned by the point on the helter shelter at which yarn or wool supplies had been contracted. It was at this point that McNairn decided to get rid of his three best qualified executives.

Tom Donaldson, the Cost Accountant, who had set up a working job costing system, and who was working on full blooded standard costing system, was the first to be sacked, for no identifiable reason. Next to go was Paterson, the financial accountant, who resigned but whose resignation would, in today's climate be perhaps classifiable as constructive dismissal in the light of his resignation statement to McNairn, "The Board of a limited Company, Mr. McNairn, is a public body and should not be deceived". His time at Selkirk had not been the easiest of his life, not helped by his failure to execute to McNairn's satisfaction his first commission, which was to justify in figures the benefit of the MD having a Rolls Royce instead of a Humber! Frank Schofield, the Mill Manager, was sacked in April after a long running battle with McNairn on the most cost effective homes for Capital expenditure. Among other projects, having just got the immense mending problem created by the crepe tweed cloths under control, he favoured a modest extension to the mending department while McNairn sought a new pattern weaving shed. Their ongoing disagreement was not helped by Schofield sending a written statement of his views to the Chairman the day before a Board meeting, but the immediate cause of his dismissal in April 1952 is not known.

All this threw the information systems into immediate chaos at time when correct and up to date data was vital to decision making. George Lindores was appointed Mill Manager, and did the job supremely well for many many years, but he did not have Frank Schofield's deep technical understanding so his contribution was less broad based. Following these departures the Managing Director's reports did not have the skeleton of figures on which to hang the flesh, and the minutes of meetings, written by McNairn expand to fill the void with generalisations. The wool market was de-established, the product was being radically changed, the database had gone missing, the machinery was not wholly suitable and the overdraft was rising. It was all getting a bit "flaky".

The data vacuum persisted far longer than either it should have or one would have expected, given the management problems which it caused. Henry Towns, the Company Secretary, simply did not have the training to respond effectively to the problems and it was not until October 1953 that, at Philip Robinson's suggestion, Lewis Ltd.'s Chief Accountant, David Drysdale, spent a few days at the Mill. He left behind the outlines of a basic set of monthly figures which could be put together from data already in existence. These were available from November and the minute of the meeting of 8 December records McNairn's appreciation of the help given; "His considerable help had made possible the provision from 1 November of the figures which the Managing Director regarded as essential to the efficient running of the Business". David Drysdale also alerted the Board to the urgent need for a properly qualified accountant at Gardiner.

Nor was America the only country with agency problems. In June McNairn reported that Continental markets had not been very lucrative in the past few years, a state of affairs which he attributed to poor representation and he proposed changes coupled with a visit by himself and Miss Sheila McIvor, his personal secretary who was fluent in French and German, to Switzerland to show the Collection.

A new agent was appointed in Italy and the women's wear Collection was sent to him with instructions also to survey the Men's wear field. Sweden was considered not to warrant a visit but the Collection was sent to the agent there with suitable exhortations. In November McNairn decided to transfer Miss McIvor from secretarial work to overseas selling, particularly in Switzerland, Italy and possibly Scandinavia with an increase in her salary and a commission on sales. Whether this arrangement ever became effective is open to doubt, but the salary increase was no doubt welcome.

The finance raised by Glasgow Industrial Finance Ltd., had the effect of removing the twin problems of yarn stock and the overdraft from the records of meetings right through to November 1952, when yarn stocking policy once again raised it's head. Up to the 1951 slump in wool prices brought about by speculation attendant on the Korean war the policy had been to buy some yarn ahead of the selling season so that it would be in hand to support orders as they came in. The slump had caused this policy to be changed, with yarn only being bought against cloth orders, which had an adverse effect on cloth delivery dates at a time when, the buyer's market now being fully developed, the lead times sought by customers were shortening. The minutes of the November meeting record a completely open ended licence to buy yarn ahead of actual requirements. "The Board considered the problem and decided that, in principle, the Company should buy yarn ahead of each season in quantity sufficient to enable it to make its deliveries of cloth on the dates required". Implementation of this policy was delayed until the London wools sales opened on 26 November so that the trend of the market might be better assessed.

The financial damage was not long in coming, for on 27 January 1953 it is recorded that "owing to the heavy seasonal intake of yarn now and for the next two or three months the financing of these deliveries is likely to bring the overdraft into the region of £65,000 or thereby". An overdraft of this size required an approach to the bank for accommodation, which was forthcoming with a limit of that sum agreed to the end of July. But before that the yarn stock, and other factors, were pursuing the overdraft upwards beyond that limit which is hardly surprising in the absence of a truly effective management information system. On 18 June, at what must have been a lively meeting, the Board recognised the need for more capital and agreed to approach the bank for a new limit of £80,000, the maximum permitted by the articles without reference to the ordinary and preference shareholders. The reasons given for the cash shortage were threefold. First the output limitations of difficult cloths imposed by the "number of effective looms". Secondly, delay by customers in taking in goods and thirdly the heavy intake of expensive cashmere yarns for the USA.

Four new looms had been delivered in January but McNairn, despite the current financial problems, reactivated his January suggestion that the 27 obsolete Dobcross looms be replaced over two years by 14 Hattersley Standards at a cost of £15,400, citing the shortage of suitable looms for the current product mix, coupled with the serious under recovery of overheads flowing from that shortage, as justification. In the course of the argument, unbeknownst to the Directors at the time, the dangers of lack of data shine like a lighthouse. Reference was made to a probable under recovery of overheads of £19,000 for the financial year "which would reduce the estimated profit for the year to about £20,000. At this point the first half year was not completed, and it was to turn out much worse than expected. Not surprisingly the Board, after referring to the increases in stocks of all kinds which would flow from such an increase in capacity, put back consideration of the proposition until after the end of the financial year.

By this time relations between the Board and their Managing Director were less than cordial which was most unfortunate for both the Company and McNairn. The Board were fed up with what seemed to them to be McNairn's money guzzling plans in the face of a deteriorating cash situation, and were not happy with the quality of the information on which they had to base their decisions, while McNairn, though without the backing of a numerate executive to prop up his arguments, was, in fact, absolutely right. Unless the 27 obsolete looms could be brought into production profit in the market conditions of the day, was not possible. Research, not many years later, when conditions were much the same, showed that in terms of size there was no place for a weaving mill with between 20 and 60 looms. Below 20 the very specialist design based producer of limited runs could live, and above 60 there was a place for the quality producer for the middle range of garment maker, but in between a mill fell into a vacuum. A real clash of personalities had grown up, a clash between

McNairn's expansionist, enthusiastic approach to the conduct of affairs and the corporate mismatch of the other shareholders, one a very successful retail group and the other a truly struggling textile business which couldn't have put up money even if it had wanted to. This combination of major shareholders virtually guaranteed that Gardiner had to reconcile itself to living as best it could in an undercapitalised condition, a constraint which suited McNairn's temperament not at all.

The only alternative to replacing the Dobcross looms was to design something which could be made on them, McNairn recognised this alternative solution to the problem and by the next Board meeting on 13 October he had cracked it; a very good thing, for his reports to that meeting were on the dark side of gloomy. He had to report, the half year's accounts being to hand and showing a profit of only £2,000, that the under recovery of overheads was £17,500, not the £10,000 expected, largely as a result of additional expenses in the American venture. Exploration of the prospects for the future settled on a probable year's profit of £7,000 a figure plucked out of the air in the absence of a proper database. Possible economies were bandied about, but without a conscious change of policy none of these were realistic. However, McNairn was able to reveal his solution to the Dobcross problem, wool and mohair raised cloths, traditionally made in Yorkshire, cloths where the finishing processes covered up inaccurate weaving. Described by McNairn as a technical advance by the Company he could not, with the overdraft pushing against it's limit, but end his presentation with reference to money. "The problem, however, was the financing of this greatly increased capacity, and the Managing Director undertook to look into this matter very carefully".

The year 1953, which in terms of the financial year to end January 1954 was to show a profit of £2,341, against a loss in the previous year of £21,351, saw it's last Board meeting on 8 December and the minutes of that meeting summarise so well the state of the game that large parts of them are worth reproducing verbatim.

" **Managing Director's report for the month of October 1953.**

a)   **Finance**

The report was prepared for a meeting on 24 November and the overdraft figure was therefore quoted at 17 November at

£54,952, with a note of anticipated overdraft at 30 November of £63,000.

It was confirmed at the meeting that this figure had not been exceeded, and it was noted that this was £17,000 less than the anticipated peak of £80,000.

The Managing Director indicated further that from the new weekly figures available to him from 1 November it appeared that with a cloth deliveries commitment of £81,000 for delivery in November / December, for which payment should be received between 15 December 1953 and 31 January 1954, against a small yarn commitment of £24,000 and expenses of £18,000 the overdraft should be substantially reduced.

He reported further that yarn purchasing and been strictly curtailed, and even where absolutely necessary had been delayed until the last moment on account of an anticipated fall in wool prices.

He stated, however, that if the plant were to be kept running at a level to recover the full amount of overheads in the financial year commencing 1 February 1954, we must order yarn regardless of possible falls in wool prices.

He indicated further that yarn had been bought ahead for a limited period, and that, having placed these yarn orders he was now pressing the spinner for almost immediate delivery.

For the winter season 1954 the prospects seem to be good. The cloths, both of the Company's Tweed types and the new continuity Pile fabrics had been well placed with merchants and with garment manufacturers. The probability was that as soon as customers started showing their patterns, or had made up their model garments, we would have substantial orders. If this was so we would require to take in yarn very quickly to execute these orders.

He reported further that the new figures available from 1 November had revealed a large apparently unnecessary, cushion of certain yarn qualities in stock, and he stated that a new programme had been instituted whereby only yarn which was immediately consumable would be taken in.

In parallel with this he reported that it would be some months before the Company was able to reduce substantially the volume of unnecessary yarn in stock as it could be used only with new yarns of the same kind coming in from the spinner.

As a result of this overlap in clearing yarn, which had accumulated unnecessarily, with the taking in of new yarn for the pile fabrics, and for already established cloths using different colours or counts of yarn, we would have a high overdraft, level again in April and May, but that after these dates the financial position should become much easier.

He pointed out that the maximum permissible overdraft without reference to the Preference Shareholders was £80,000".

There followed sales, orders and production figures, though no stock figure is presented. The comments on unnecessary stock in the main paragraph are supported by a sales orders figure of £126,223 representing a yarn content of £75,733 and a yarn on order figure of an almost exactly matching £75,332. Normal consumption was not going to dent the yarn stock at all.

Only minor changes in machinery were made in this period, 4 new Hattersley Standard looms were delivered, and a Moser raising machine was bought and minor items in the winding departments were acquired. The teazle raising machines and cropping machines to finish the pile fabrics being made to use the Dobcross looms had yet to be bought. The new building to house the pattern department and the designers was in service by the end of 1952 having it's own looms, winding and twisting machinery, stake warping and it's own yarn store.

Dividends of 5% had been paid but in the year to 31/1/53 in which a loss of £21,351 was sustained no payment was made and the previous years dividend was to be the last for some time. The Preference dividends and redemptions were kept up to date, though at 31/1/53 the profit available for those purposes had fallen to a mere £3,853, insufficient to make the 1954 payment should nothing further be earned unless certain reserves were brought into the profit and loss account. In the year of loss the Board made a payment of £1000 to the MD though there was no entitlement to commission in his agreement in the event of a loss. Since in recent years commission on profits had more than doubled his salary a complete absence of bonus would he been pretty disastrous to him, a fact which the Board recognised on that occasion.

During 1953 a new designer was appointed. The outside applicants for the post having all been found inadequate McNairn promoted George Oliver, a young Selkirk man who was working as a warper at Tweed Mills. He had studied design at the Scottish Woollen Technical College in Galashiels and very successfully made the change from blue to white collar.

At the meeting on 8 December 1953 it was decided that Henry Towns should retire, and that Ian Jackson, ACA should become Company Secretary, both on 1 January 1954.

Chapter Eight

Home to Roost 1953/1955

The first Board meeting which I, having become the Company's Secretary and Accountant on 1 January 1954, attended is very well documented. I had not learned the technique of writing succinct minutes, so the record of the meeting of 2 February is far more detailed than it should have been and shows very clearly the differences in attitudes between not only McNairn and his colleagues, but also the differences between Watson, Campbell and Robinson. The different pressures on and backgrounds of the Board members made it a less than harmonious gathering.

Sweetness and light prevailed for a few minutes while Hugh Watson congratulated James Campbell on his knighthood, and went on to welcome me as Secretary. But when the Chairman initiated a discussion of the Managing Director's report the fur began to fly.

To soften up the Board the opening sentence of the MD's report read: "The prospects for the Company are more favourable now than they have been at any time in the last three years". This was in fact true, but might perhaps lead the Board to read the rest of the report with somewhat jaundiced eyes.

After guessing that Gardiner should show a profit "however small" (in the event £2341) for the year to 31 January 1954 it went on to review the Company's situation. £90,000 had been sold in the new cloths in the American market, a lot had been learned and substantial yarn stocks were in place to support this trade. The raised fabrics, which could bring back into service the suspect Dobcross looms had been well received, but these too were, of necessity, supported by yarn stocks but to a lesser extent, for the cloths were for piece dyeing, so only grey yarn need be stocked. Cropping machines had been modified and second-hand teazle gigs purchased to provide a complete finishing plant for these cloths at a total cost of £2000. The new "Peach Bloom" fabrics had come out well in comparisons with the French cloths which they were emulating.

There was substantial spring repeat business on the books, and America, after an almost non existent Winter 1953 season, had come away well for spring with orders for 250 pieces, half of them cashmere, on the books. All true, all encouraging, none of it likely to upset the Board.

But the report then, quite properly, rehearses the argument that 46 Hattersley looms cannot earn a proper profit, that cannibalising 7 Dobcross looms has allowed 20 to be brought into service to make the pile cloths. The yarn stock implications are left in the air while the report goes on to outline the sorry state into which the yarn stock has got. The failed USA winter season had left a hangover of £51,000 worth of yarn, of which the current order book would consume a mere £15,000 were it's needs entirely covered in the stock. Having disclosed that piece of unpalatable information the report then acknowledges that there is "a large unnecessary cushion of each yarn quality in stock", and goes on to analyse some of the causes yarn by yarn and outlines a programme of yarn substitution which, whilst working away the surplus and reducing the intake will involve stock losses.

Turning to finance the report identifies the reasons for the year end overdraft forecast having moved, since 8 December, from £24,000 to £65,000. They were an increase in trading activity, increased yarn intake, shortfall in actual cloth deliveries and payment delays by customers. This analysis led into the development of a perfectly correct argument that if proper volume were to be achieved this level of overdraft must apply for some time, and the profit figure plucked out of the air was £3000 a month, with the risk that any delay in payment by customers, or shortfall in deliveries leading to reduced income, would be liable to push the overdraft up to £95,000 requiring the Preference shareholder's consent, which McNairn proposed be sought.

The rest of the report comprised figures for sales, orders in hand, yarn intake, production, yarn orders outstanding and employees and wages, followed by reports of minor asset purchases and disposals.

The Board argument centred round the proposal that the overdraft limit be increased, which proposition was rejected with the Managing Director being instructed to achieve the maximum possible profit and turnover within the existing resources, and reducing overheads by whatever means possible. The discussion was wide ranging, with James Campbell questioning the past policy of modernisation and expansion and Philip Robinson hammering the classic retailer's point of view, reduction of stocks and increase of margins. Hugh Watson stuck to the work within resources line, which really was the only option while McNairn fought for more money while he was sorting out the stock problems. The whole sales/capacity/overhead recovery/resources equation was well chewed over and the desirability of abandoning the American market for the sake of reducing the overdraft was thought of, as was the possibility that the Company was perhaps not worth keeping going. At the end of it all nothing, except the undesirability of approaching the Preference shareholders and the payment of their dividend, was really decided.

McNairn, and his colleagues, were in an impossible position, it must have been clear to all that the Company was undercapitalised, and that that situation was exacerbated by errors in balancing yarn intake against consumption, the errors almost certainly flowing from a combination of over optimism and lack of proper statistics. Attitudes and positions were hardening.

The following meeting, on 3 March 1954, was largely devoted to detail. Freddie Stark of the firm Frederica, offered large orders in return for extended credit. The Board's response was consistent with the "work within your resources" directive. Possible overhead savings were talked of, with McNairn being restrained from over heavy use of the axe. The significant discussion concerned yarn stocks, where the Board had to argue against most of McNairn's over enthusiastic proposals which would have seriously impaired the Company's viability. They did, however, agree to withdraw from the 100% cashmere market and use away or dispose of the supporting yarn stock.

At my prompting the Bad Debts, Yarn Stock and Tax equalisation reserves were rescued and returned to the Profit and Loss account. The Profit and Loss balance was just adequate to cover the Preference redemption and dividend for the year but with things the way they looked some insurance was needed.

On 16 March the lottery for Preference redemption as held and the accounts for the year just ended approved for submission to the members. The AGM and the next Board meeting were set for 11 May.

McNairn's response to the Boards rejection of his expansionist plans on 2 February was either irrational, born of sheer frustration and consequent upon his colleagues' apparent refusal to recognise the unbreakable connection between size, volume and profitability, or constituted a very un-subtle form of attempted blackmail. Between 2 February and 11 May he set in train a sweeping set of policy changes and reforms which went far beyond the needs of the simple directive "work within your resources". Though he had on site a Chartered Accountant who, despite his severe lack of experience of the industry could none the less provide the building blocks for "what if" equations the MD continued to do his own arithmetic, which was oriented toward the justification of his actions and, regrettably, the development of some inaccurate and misleading projections. There can be no doubt that having disposed of a good management team he had decided to operate as a one man band and quite probably regarded me as a quisling or similar animal.

Typical of his attitude was his view of costing, or cost allocation. Virtually since he had been at Tweed Mills cloth costings had comprised, firstly a calculation of the cost of the yarn content of a single piece warped 56 ells long and finishing somewhere between 58 and 64 yards, depending on cloth type, by the time it landed on the warehouse table. To that were added the direct wage costs. Some of these were derived from piece work rates, and some from calculations based on wage rates and estimated output. Some of these came out as rates per pound of yarn processed, some attached to the number of picks per inch at the weaving stage and some per yard of cloth processed.

With the limited statistics available, and in the absence of any formal budget for output or expenses, only a limited slice of the productive wages bill could be applied in this way, so the balance of such wages and all the other expenses incurred by the Company, including the costs of the pattern department, were lumped together as "overheads", and applied to the costing after the costs to the pre-overhead point had been reduced to a per yard figure on the basis of the predicted yield, in yards, of 56 ells warped. In 1954 the figure was 1 °d per pick.

When I had got the January 1954 account out of the way I spent months devising a full blooded system of departmental budgets and standard costing, just as Paterson had done before me and been forced out for his pains. In due course I presented it to McNairn in the form of a substantial paper, which he took away and returned days later with the simple comment: "Interesting, We will continue to cost on a penny halfpenny a pick". He clearly did not want to face the fundamental policy problems which a change of basis would raise. Which was a pity, for the Board, presented with a proper analysis of the effects of their policies, might well have come to conclusions which might have avoided the pain which was imminent.

Between February and May, McNairn attempted to do much more than work within his resources. Having been refused an extra £15,000 of overdraft he set about reducing it on a permanent basis, below the old limit of £65,000, taking violent actions designed, according to him, to produce a profit off a turnover of £350,000, some £130,000 less than that achieved in 1952. The steps taken included the discontinuance of many cloth qualities and their associated yarns. The rigorous limitation of yarn intake REGARDLESS OF PIECE DELIVERY REQUIREMENT. Closing down the Dobcross shed of 20 looms and selling off all cloth stocks regardless of price. Quite apart from losses on selling off piece stocks discontinuance of yarn qualities inevitably brought yarn stock losses in it's train. The designing and pattern making section was curtailed by the transfer of it's broad looms to main production. Some production employees were paid off and the shedding of a designer was mooted. So much for "We're not selling cloth, we're selling fashion ingenuity". Advertising was discontinued. All this, according to the McNairn arithmetic, to achieve an annual profit of £7,000.

These activities were part of the MD's report to the Board on 11 May 1954, with a telling paragraph reading "In achieving this contraction we will have made a stock loss on yarns and on pieces. We have prejudiced delivery promises, though by great effort in production not too seriously. We have lost something in prestige locally by closing down a weaving shed, and with spinners by seeking extended credit when the strength of our shareholders is known to be substantial. My personal loss of prestige and my embarrassment have been considerable". Another paragraph read "As a basis of trading it is in my view most unsatisfactory, and it can continue with the curtailed pattern making only so long as the Gardiner reputation for style and novelty remains, which would be about three seasons".

In the concluding paragraphs of the May report to the Board, McNairn posed two alternative courses of action. The first was "To amend the decision of 2 February and restart the Dobcross shed" or to send out the pile cloths to be woven on commission. Reference is made to the stock costs involved, and the effect on profit.

The second was to buy 8 new Hattersley looms. Again there are figures for the effect on profits, for capital costs and for the cost of yarn stock support. Nowhere is there a suggestion as to the source of these funds.

It all went down like a lead balloon. The Chairman opened the discussion and "expressed disappointment on behalf of himself and the other members of the Board that the Managing Director had not allowed the new basis of trading decided on 2 February 1954 to run for at least nine months before suggesting these amendments". To which McNairn replied that he suggested the changes because he considered that the policy was too tight and exposed the Company to considerable risk of loss.

After a somewhat meaningless discussion the Board allowed that if the overdraft did in fact fall to £15,000 below the levels obtaining in late 1953 then work might be sent out on commission. But, profit forecast of only £7000 notwithstanding, the February policy stood and would prevail throughout the financial year beginning 1 February 1955, even though, with commission weaving, a mere £11,000 of profit could be hoped for.

As the half year to July 1954 progressed the frictions between McNairn and the others, and between the principal shareholders themselves, deepened and the possibility of reconciliation grew more remote. The

20 July meeting received a report showing a likely five months loss of £8,400 and projected, on questionable grounds, a profit of nine thousand pounds for the rest of the year. The first half loss was, therefore, a surprise, for a profit of £1700 for the half year had been projected, and the Board did not, therefore, accept as probable the forecast for the closing seven months. The difference was attributed to excessive under recovery of overheads and unexpected stock losses with a overriding caveat on both the result and the forecast to the effect that much depended on the market mix of margins which, in the projections, was as yet unknown.

By this time meetings were being held in the Chairman's office in St. Andew's Square, which gave the principal shareholders the opportunity to meet beforehand to decide tactics, which allowed the Chairman to set the background in his opening remarks and on this occasion he threw a bombshell into the room, saying that the divergence of views between the MD and his colleagues on policy had created an impasse, with the MD requiring more capital and the others unwilling to provide it. There seemed to him to be two possible courses of action. Either the MD worked within the existing resources, or he arranged for the UTR and Selfridge shareholdings to be purchased by someone else. McNairn's reaction to this idea is not minuted and he seems to have played little part in the discussions at that meeting.

Philip Robinson stated that the Chairmen of the two companies had agreed to act in concert in any sale. Sir James Campbell underlined McNairn's dilemma by pointing out that not only would he be required to find the purchase money but, on his own arguments, another £40,000 or so of capital as well, and went on to say that he thought UTR might accept fifteen shillings a share, which valued their holding at £27,750 and opined that a profit of £15,000 or £16,000 should be possible for next year. No source for this estimate was hinted at. Philip Robinson declared himself unauthorised to contemplate selling for less than par and it was agreed that any relevant suggestions should be put to the Gardiner Board, and the Chairmen of the investing Companies should not act without a recommendation from that body.

The turnover limit, which had never in fact been Board policy, was removed, and replaced by a simple instruction to work within the existing resources. Payment of the Preference Dividend due on 31 July was authorised, even though the revenue reserves available, were the current projections correct, would only be £7,600.

The accounts for the half year to 31 July 1954 duly produced and showed a loss of £21,071, two and a half times the May forecast. The regular September meeting had been set for 14 September, but a special meeting was held on the first of the month in Glasgow, so that the Chairmen of the investing Companies could make their views known. Selfridges had a store in Glasgow, and UTR were based only a few miles out of town. To begin with the Board alone considered the accounts and heard McNairn's brief explanation of the divergence between forecast and outturn. It was agreed that to achieve a result for the year which would allow the Company to meet it's obligations to the Preference Shareholders at the end of the year a turnover greater than presently envisaged would have to be achieved and the MD was authorised to work right up to the agreed £80,000 Bank facility.

Then, Mr. S. H. Leake, Chairman of Selfridges, and Mr. W. M. Calderhead, Managing Director of UTR joined the meeting.

Mr. Leake, a dry as dust, incisive person, batted first and opened by saying that the steps proposed for setting things right were merely palliatives and that "it seemed possible that the Company was under capitalised", going on to suggest that the Board consider whether the plant was satisfactory for carrying on trade in present conditions. If, after consideration, new plant were found to be necessary, he suggested the following:–

That the 20,000 undenominated shares be converted into four or five percent Preferred Ordinary Shares and taken up equally by the two principal shareholders, and that if UTR were unable or unwilling to take up their part he would consult his own Board to decide what action Selfridges should take. He went on to suggest that this £20,000 would automatically raise the permissible overdraft limit and with touching faith concluded that the capital made available would therefore be not £20,000 but £40,000. This sum would both buy and finance the operation of 16 new Hattersley looms.

Having thrown this brick into the pond he said that he was willing to listen to any other suggestions!

No other suggestion was forthcoming except that Mr. Calderhead offered the thought that if the Company stuck to high novelty fabrics bearing large margins a decent profit could be made on the existing complement of looms. Sir James Campbell was deputed to tell Glasgow Industrial Finance of the half year's result, but not the reasons for it, and to inform them of the Board's short and long term plans to improve the Company's position. Some task, since the Board with it's conflicts really had no coherent plan.

The detailed consideration of the half year's account was at the meeting on 14 September in Leeds.

Yet again under recovery of overheads was advanced as the reason for the loss exceeding predictions by so wide a margin, but this was more a political explanation than a factual one. Under recovery there certainly was, with the Dobcross shed out of commission, but the stock losses resulting from the twists and turns in stockholding and product policy were the real reason. In normal conditions at that time the yarn content of sales averaged between 60 and 65 percent of the selling price of cloth. In the corresponding half year the figure was 62.6%. In the half year under review it was 77%, so either prices did not include proper margins or substantial stock losses had been incurred.

Though not put forward as one of the causes of the loss there can be no doubt that it was a substantial element, indirectly acknowledged by McNairn in his refusal to forecast a second half result. In his rambling consideration of possibilities he suggested that there might in the second half be a benefit from yarn in the yarnstore gaining condition, or taking up moisture and therefore gaining weight which, by implication it had lost in the first half. Since in properly conditioned yarn water is some 20% of the weight, a long hot summer might make a difference but experience shows that yarn is almost never up to condition and that the actual condition varies little unless a conscious effort is made to put it in, by steaming or some other method.

The results having been swept under the carpet the Board turned to the future and approved substantial price increases to offset the reduced productive base, which increases, because of the seasonal timing of offering, could not be effective in the current year. Philip Robinson went on to that if, by July 1955, they had proved effective, fresh capital should be sought if it could increase profits. James Campbell reacted in an uncharacteristic way by doubting that Mr. Leake's proposals for increased capital were dependent on prior profitability. This point of view was supported by the MD with the caveat that as, without the money, the plant was in fact 46 looms, not 62, then 46 had to carry the overhead burden. Robinson hedged for all he was worth, clearly riding under different orders from those which prevailed a fortnight earlier.

The argument rolled on, but with the prosperous shareholder (Selfridges) hedging it's bets, UTR was not going to force the pace of capital injection, and no conclusion was reached. It was as much the differences between the principal investors as the rift between them and McNairn which caused the zig zag course of policy, or lack of policy, at this time. A lack of purpose which could easily have brought the Company down.

Mr. Robinson's suggestion that a non working director knowledgeable in yarn be sought was countered by the MD saying that his new personal assistant, Henry Colin Brown, had been appointed with just this in mind, a tale slightly economical with the truth. They had met at David Maxwell's wedding, had got on and, Colin being disgruntled in his job at Laidlaw and Fairgrieve because of dynastic problems, had agreed to move to Gardiner on a very vague job specification!

Yet another change of policy. Sir James was told not, for the present, to advise Glasgow Industrial Finance of the half year's result.

Relations were now at a terminal stage, and that was the last meeting at which McNairn made any noticeable contribution, though he continued to design the collection, travel, run the mill and carry on the functions of Managing Director. He was not present, through illness, at the December meeting, at which a small profit for the second half was forecast. In the event it was £877, giving a loss for the year as a whole of £20,194. Hugh Watson agreed to collaborate with the

Chairman of UTR in informing Glasgow Industrial Finance of the position.

The January meeting discussed policy not at all, reviewed a claim against a spinner and instructed the Secretary to consult with Hugh Watson and Ewen McNairn in the composition of a letter advising the Preference Shareholders that their January dividend would not be forthcoming.

The real business of the meeting was conducted in the absence of the Secretary, and became a minute in a secret, locked minute book which was, for a time, in my possession but without a key. It recorded the agreement that McNairn should resign at the next meeting, and as far as I know that is the only minute in that book!

The next meeting, on 3 March considered the year's accounts to January 1955. The capital and general reserves were drawn back to the Profit and Loss account and it was recognised that Preference redemptions and the July dividend were casualties.

"Mr. R. M. S. McNairn, having completed the designing of the spring 1956 collections, indicated that, with the Board's approval, he would relinquish his office in accordance with the arrangements made at the meeting of 25 January, when the formal business of this meeting was completed". A well kept secret. The lads left carrying the ball, H. C. Brown, who was made General Manager later in the meeting, and Ian Jackson, the Company Secretary, only then discovered "the arrangements" made in January.

McNairn signed transfers of his shares in favour of the principal shareholders, making them equal owners of the whole ordinary capital with 40,000 shares each, and left the meeting and the Company which he had run, mostly very successfully, for a dozen years. The Chairman wished him well on behalf of the Board.

The details of who to tell and how, who should sign cheques and who should advise the customers were sorted out before the usual discussion of yarn buying policy, the usual reference to working within resources and before the relative merits of loose wool and package dyed yarn were talked of. The purchase of two second hand narrow fast looms for pattern making and the sale of two Dobcross looms were authorised.

So ended the McNairn era. Profits had moved from £5725 in 1946 (31 January) to a peak of £36,268 in 1952, before crumbling into losses. Net worth had risen from £37,404 to £83,077 having peaked at £129,572 before losses took their toll. For the times it was a very acceptable track record. If only . . . .

Chapter Nine

New Team, False Start 1955/1959

The Board in their wisdom, perhaps having had enough of Managing Directors for the time being, told the "lads", H. Colin Brown, the General Manager, and Ian Jackson, the Secretary and Accountant, to carry the business on "within the resources available".

Colin Brown was a Borderer born and bred. In the nineteenth century his family had owned the Galashiels weaving firm of Brown Brothers and had sold it on to others in the early part of the twentieth century. Colin himself was born in 1924 and after education at St. Mary's in Melrose and Loretto in Musselburgh went to the R.N.V.R. from school. After training at HMS Ganges he joined the cruiser "Sheffield" as an ordinary seaman and in 1942 and 1943 escorted Russian convoys, emerging with damage no worse than frostbite. After officer training he joined the destroyer "Icarus" as a sub lieutenant, leaving her as a lieutenant to go minesweeping in 1945 and 1946 until being demobbed in that year. Colin remained in the R.N.V.R. and retired with the rank of Lieutenant Commander in 1956.

Demobilisation was followed by a year at the Scottish Woollen Technical College. Spending vacations working at Crombies of Aberdeen did him good, for they sent him to Leeds University to complete his set of City and Guilds certificates and do a course in economics, after which he went to work for one of the great men of the Scottish textile industry, John Ross, Managing Director of Crombies, as his personal assistant, The years there gave him the best grounding he could possibly have in the conduct of a woollen mill in all it's facets. Then "family", the Fairgrieve family, prevailed on him to come back to Galashiels and join Russell Fairgrieve in the management of spinners Laidlaw and Fairgrieve. But Colin and Russell did not run well together in harness and, after less than a year, meeting Ewen McNairn at David Maxwell's wedding, he was persuaded to join Gardiner's.

Discussing this proposed job with his father-in-law, a prominent Leeds garment maker, Colin was given the following advice; "You can afford the experience. It will possibly last eighteen months"!

Ian Jackson came from Lancashire, though like all good Lancastrians his family had lived for years in Cheshire. Only ever had one ambition, to go to sea so, having failed to get into Dartmouth, went to the Nautical College, Pangbourne and from there, in 1943, to sea in the Blue Funnel line as a midshipman. The family did not think this was the best idea in the world.

Five years, seven voyages, three of them as Third Mate, later, family pressure won and Ian came ashore. The pressure was continuous and unsubtle and included the oft posed end of voyage question, "when are you going to get a proper job", upsetting to one who though he had one. Once ashore the problem of what, having abandoned a vocation, to do raised it's head and Ian, being entirely ignorant of shoreside and business life had little contribution to make. So he was articled and turned into a Chartered Accountant, qualifying in December 1952. This qualification was said by all to be "a good start".

In late 1953 Cooper Brothers declined to increase Ian's salary beyond £600 a year, so, being offered it, he accepted £1,250 to come to Selkirk.

This, then, was the management team which the Gardiner Board left to run the mill. Nobody guessed that the same team would still be in place thirty one years later. And, the two men being entirely different in character and background, having only the few years at sea as common ground, nobody would have bet on them staying together that long.

George Lindores was the Mill Manager, whose task, getting more complex by the day, was to "run the mill". In these nineteen fifties times that included dealing with all labour recruitment and problems short of overall wage increases, which were handled by the Scottish Woollen Trade Employers Association (SWTEA from now on). Arguments about piecework rates and the like were his responsibility, as was the routine ordering of yarn within parameters set by the General Manager. George did this job in a growing business extremely well until he retired in 1978.

The department known as the sample room was managed by Walter Turnbull whose title, surprisingly, was Selling Office Manager, though his involvement with selling went no further than sending out pattern collections to the agents. The cutting up of pattern ranges and the bunching of those selected by the designers for showing was sample room work. So too was the sending out of confirmations of orders, each illustrated with small clippings of the patterns ordered, and the passing of copies of the orders to the Mill Office for analysis of yarn content and yarn ordering. Walter kept the pattern books in which a clip from each design made was recorded, and the seasonal books in which clips of all items sold were recorded, with quantities, by customer. He also kept the order book in which orders were listed and totalled weekly.

The Mill office was the province of Alec Wyse, who analysed orders into yarn quantities by quality and colour and listed them on a "white sheet", one for each item. A laborious task which provided incomplete information to George Lindores for his yarn ordering, for the stock was kept by Jack McDonald on a card index which did not include a total of the yarn involved in outstanding orders. Small wonder that the stock went wrong from time to time.

The Yarn Store was the province of George Watson, a great character and one of nature's gentlemen. A Galashiels man he came to Gardiner from William Bliss in Chipping Norton, where he had been a Minor Counties cricketer, so he was the backbone of the mill cricket team. A natural sportsman he was a good golfer and took up fishing with great success in his later years. A very tall man he was unfortunately dogged by ill health, though from his demeanour you would never have guessed it.

John Kemp was foreman of the weaving shed, responsible for all aspects of the weaving operation from seeing that all the looms were filled to the quality of the product as it left them. He is perhaps best remembered for his habit, when work was tight and looms in danger of standing empty, of rushing across the yard to George Lindores to announce "we've crashed". It was usually, given his skill, an incorrect forecast.

Finishing, wet and dry, was John Knox's job. Getting raw cloth from the weaving shed he saw that May Reid and her girls mended, or darned, it properly, then saw to the milling and scouring of the cloth, to its cropping to shave off the surplus fibre and then to the "blowing" or steam pressing before it passed across a table for final inspection before crossing the yard to the Warehouse for final final inspection, rolling up in one package form or another and wrapping up for despatch. And then the Warehouse would arrange for the cloth's transport to the customer. Masterminding the Warehouse at this time was Charlie Brodie. We were not then to know that in later years his son would be our Bank Manager. In those days packing cloth was not just a matter of slipping it into polythene tube and sealing the ends. Home trade fabric was wrapped in real Kraft paper then baled in hessian, whilst exports were either so baled or baled then put in packing cases made to order by the mill joiner. Another practice of the time which survived a few years before it's controversial abandonment was "fringing". The weaving process in a conventional loom leaves loops of yarn along the selvedge where a shuttle has lain idle whilst others pick before coming back into pattern. These loops were manually removed before the cloth went to the warehouse, a time consuming operation.

The Head Designer at the time was Alex Linton, nearing his retirement. Nobody knew the colour or quantity of Alex's hair for he wore a cap which he never, ever, removed. A real Selkirk worthy who visiting Americans on designing expeditions must have found hard to understand. He was supported by two much younger men, both in their twenties, Janus Chasczcynski, the son of the Colonel and his wife who ran the mill shop, and George Oliver, a young man who, working as a warper but trained in design by the Scottish Woollen Technical College in Galashiels, had been spotted by McNairn and moved into the design office. The designers were supported by clerks who translated the designs from their ex-designer form into weaving tickets for the pattern shop.

The pattern shop was presided over by Will Scott, a bachelor, another Selkirk worthy. He was found to be colour blind, but judging by his performance no one would have suspected it. In later years Will came into some money and became interested in motor cars. He started with a Messerschmidt fore and aft two seater, but later, on the occasion that Miss World made a tour of the Borders and had to be transferred from Pringle's in Hawick to Gardiners in Selkirk the job was given to Will, not because he was a bachelor but because he had the best car in the mill, a Jaguar, no less.

Jessie Galbraith was the cashier, kept the books and ran the general office which at that time had about four other people in it. She also did the confidential typing, for we felt that a private secretary was something we could ill afford and Sheila MacIvor took her departure.

The year to 31 January 1956, whose foundations had been laid by McNairn, fulfilled his optimistic predictions and yielded a profit of £30,810. In later years we came to regard this as almost unfortunate, for it led us to believe that present policies and products were adequate for the foreseeable future and it took five years for us to be disabused of this idea and to formulate an alternative strategy, of which more later. The following year to January 1957 produced £17,922, the next £2,598 and the next, to January 1959, a loss of eighty eight pounds. A revival in 1960 showed a profit of £14,690. All in all a progression which did nobody any credit. The chapter heading of "False Start" could equally well have been "The Years of Lost Opportunity".

With no Managing Director and two unknown quantities, Colin and Ian, running the place, it is not surprising that the Board took a close interest in day to day affairs. We managed to avoid the yarn stock problems which had bedevilled McNairn partly because Colin had his feet more firmly on the ground and was not prone to invent new cloths and stock yarn to support their unproven worth and partly because management information improved. Quite soon the standard costing system, on which most of the work had already been done, was working and the gaps in the yarn stock information had been plugged. Philip Robinson, of Selfridges, was the most active Board member, taking it upon himself to speak to or interview anyone, agents, overseas prospects or whoever, who passed through London. He had a real salesman's outlook and was of the view that if we could get the sales force right and motivated everything else would fall into place. Odd, for that viewpoint was in conflict with his retailer's position that the product had to be right if it were to be stocked. Rightly he had a high opinion of himself and his capabilities. The top tax rates at that time were over ninety percent and Philip, a very high earner, was prone to tell us frequently that his Rolls had been bought out of taxed income, which tale we always took with a pinch of salt.

Sir James Campbell was, as his background of declining businesses would suggest, the epitome of caution and it may well be that his views held back the essential re-orientation, or in today's business-speak "repositioning" of the Company. His reluctance to use pricing policy as a marketing tool undoubtedly, in the light of events in the sixties, delayed Gardiner's revival.

Hugh Watson was the perfect Chairman. With a mind like a razor he cut through the flim flam and minimised the conflicts between Robinson and Campbell whilst keeping a balance between the aspirations of the managers and the interference of the Board. He was knighted in the 1957 New Year's honours list and well deserved it.

Shortly after McNairn's departure an order for 12 new Hattersley looms which he had placed without Board Authority came to light. These were for delivery by 1958 and, Philip Robinson having failed to negotiate their cancellation, they were taken in in groups of four, the first in early 1956, 'to link with the resumption of Preference redemptions". Another early capital purchase agreed was Supercop winding machinery. Conventional looms carried weft wound on "pirns" speared on spindles within the shuttle. The Supercop concept was yarn wound on itself and withdrawn from the inside of the package. The shuttles had grooved insides to hold the package in place and, because of the lack of a wooden former the amount of yarn carried in the shuttle was more than doubled which allowed less shuttle changing and greater loom efficiency. A simple concept, but requiring accurate winding at controlled tensions if the package were to hold together as the shuttle was batted back and forth, and the problem of the collapse of the very end of the package at it's exhaustion was never really overcome.

In the following year, in July 1956, with the £30,000 profit achieved, the lads proposed their first major piece of capital expenditure, a Charlesworth Whitely tenter to replace the existing cloth drying machinery. This was approved, installed, and the machine, much modified with modern control equipment, is still in use today. At the time, at a cost of £7,000, it was quite an investment.

With Alex Linton about to retire the search for a new head designer became a priority. An early choice was Douglas Johnston, then working in the North of Scotland but he, after interview, declined to join. The resignation in July of Janus Chasczcynski injected fresh urgency into the problem which was solved by the appointment early in 1956 of Willie Riddle, then working in Ireland, to the Head Designer post. Willie stayed with Gardiner until his retirement. A little earlier a young man, Gordon Brown, had been appointed to fill the post left vacant by Janus, so the design team was up to strength again.

Design has ever been fundamental to textile success. Common representation by agents who made a substantial contribution to the styling of the collection, such as George Klein of TBT, was not without it's problems and Colin, reporting on his 1955 visit to the States observed that he had noticed similarities in the designs of Gardiner and the other mills represented by George Klein, and suggested that the Klein input must be reduced if this were to be avoided. A difficult concept, given George's character. Indeed the relationship with George Klein was, over the years, not an easy one. George was a men's wear specialist, so whenever a women's wear collection was sent to him it failed to sell. TBT had a contract giving them the exclusive right to sell the Gardiner product in the USA and George resolutely refused to let Gardiner use another agent for the women's wear side. Other concerns with this agency which surfaced from time to time, and which brought negative responses from George were the freedom enjoyed by TBT to take on other mills without any Gardiner veto, which contrasted badly with George's refusal to let Gardiner use another agent for women's wear. Various amendments to the agreement were proposed and rejected and in May 1958 the friction had reached a pitch which led to consideration of the termination of the agency, one of the periodic termination option dates being close at hand. In the event the agreement ran on. Later in that year the very large American firm of Forstmann Woollens, part of the enormous J.P. Stevens Corporation proposed that they have certain of their women's wear woollens woven in Scotland by Gardiner. This caused the TBT position to be reviewed yet again and in the end, in voluminous correspondence, George Klein was told that we intended to take on the job, and hoped that it need not upset our relationship with him. He acquiesced. The minutes record that "agreement to the operation had been indicated but in the most ungracious terms".

By this time the gilt had gone off the American men's wear gingerbread. The cashmere suiting trade had been a very important leg of the operation and this suffered a fatal blow in 1956. The Federal Trade Commission (FTC) introduced new labelling rules which, quite logically, made it improper to describe a garment as cashmere unless it had a cashmere content of more that 50%. Most of the garment makers had hitherto been misdescribing their products, the cost of cashmere being far greater than the cost of lambswool. The FTC made it plain that they were serious, so all the makers sat up and took notice. It was well known that Gardiner's cashmere suiting, having a wool worsted warp, and a weft of less that 100% cashmere, was not fifty percent cashmere, but George Klein managed to get a certificate from one of the leading testing firms that the cloth, CWB, was 51% cashmere. It was a beautiful, cashmere handling cloth, ideal for the casual suits for which it was intended, but it only had 13% of cashmere in it.

George felt sure that the Hatch report would satisfy the customers but was soon to be disabused. In November 1956 Colin was in New York, with George, negotiating the winter 1957 contract with Greiff Brothers, the largest user of CWB and they had got to the point of the order being specified and written on the order form when they demanded a certificate from Colin that the cashmere content was over 50%. They, like everyone else, knew that it wasn't and they were not prepared to rely on the test report. After long discussion, and Colin's refusal to give such a certificate, the order was torn up, and with it went, quite quickly, the rest of Gardiner's cashmere business. As an example of the misuse of labelling laws this is a classic. The product was right, but carrying the wrong label, it fell into disfavour. Some makers struggled on, correctly labelling the garments, but such is the cachet of cashmere that without the label it wouldn't sell.

Cloths with a genuine fifty percent content proved too expensive and after one more season, which included stocking yarns of compositions which would achieve the label, the cashmere business trailed away, bringing stock losses in it's wake.

In 1957 a more general blow was struck at Britain's American trade. In response to the failing domestic woollen industry protectionist legislation was forced through Congress. The protection took the form of a quota of deliveries which would, each year, be allowed in at the existing rates of duty after which, when the quota had been filled, the rate of duty would rise from 25% to 45%. Though TBT and Gardiner tried to arrange sharing of the extra duty with customers it was a mortal blow to the UK/USA trade. Effectively, as no-one knew when the quota would be filled, no-one knew the landed price of cloth any more. The American trade declined with the quota, became less important to the British industry and in due course and for many years irrelevant to Gardiner. This developing situation was both the reason for Forstmann wanting a commission weaving deal in Scotland and for Gardiner's feeling able to make a stand against George Klein.

George Klein was one of the industry's great personalities, but he really worked only for himself. When Japan began to make superb worsted fabrics he did not hesitate to take on Japanese agencies, to the detriment of his Huddersfield mills. He treated his territory as a farmer guards his land. All negotiations with him had to be against this background, and were made more difficult by his verbosity, particularly on paper. One four page missive from him included the phrase "predicated on the assumption that". Colin and I thought that he meant "if", but our lawyer Chairman assured us that it really meant "if, if"!

The years from 1955 to 1959 saw attempts to tackle the problems of representation in other export markets. In Canada Gardiner were represented by the old established firm of Sutherland and Ewing. Sutherland was very old, and Ewing relatively young and in 1955 they were regarded as very suitable agents. Despite this little progress in expanding that market was made and in late 1958, after his North American journey, Colin reported that they were attacking too narrow a customer base, one which really did not have the potential to do the market justice, and Sutherland and Ewing were urged to expand that base. A small market, the customers were anxious that their suppliers should be as exclusive as possible to them, but in Gardiner's case this had narrowed the field too much. It was to be many years later, and with a different agent, that significant business was to be done in Canada.

In West Germany Mr. Eden, our agent since the war, having failed to generate worth while business, resigned. Not until 1958 was his successor, Arthur Ernst, appointed.

In Holland Gardiner enjoyed modest success with Gunter Roosen, scion of a noted Dutch Converting house, displaced by his brother. Gunter, tall, gaunt, speaking perfect English, being married to a Scottish wife, was full of dynamism and ideas. He had other agencies and conducted a successful business but tended to get carried away. In December 1958 he proposed that he should drop all his other agencies and sell only for Gardiner in the whole of Europe on a commission basis but with a minimum commission as security. With the Common Market looming this seemed to be an idea worth exploring and a meeting was set up one Sunday in London, attended by Philip Robinson, Roosen, Colin and me. Roosen was already representing Gardiner in the Benelux Countries, and, on a temporary basis, in West Germany, and at this meeting terms of agreement for sole representation were settled, subject to both parties accepting their written form. It was a long hard day, with Gunter behaving like the perfect Dutchman and demanding far too much and Gardiners struggling to get an arrangement which would be of benefit to the Mill. As Philip Robinson commented when Gunter had left "'T' was ever the fault of the Dutch, to give too little and ask too much". Colin and I thought the same of retailers, but held our peace.

A week later, after meeting with Sir Hugh, a letter was sent setting out the terms agreed. Early in January Roosen decided that he did not want to proceed and we were left looking, once more, for a German Agent.

On the design front Gordon Brown left us to do his own thing, and progressed through managing a Galashiels mill to being heavily involved with a Dutch fancy spinner, through property and on to tax exile in Jersey. Derek Garret joined as a men's wear designer, a Hawick man, he came to us from a Yorkshire mill and remained for many years. Alex Linton retired in May 1959. In April 1957 an agreement was made with M. Fred Carlin, a Paris design consultant, to help us with our styling for two years, with a view to giving our own designers a new perspective and vitalise our styling. The benefits of such arrangements are unquantifiable, but there is no evidence that it did much good. Nor did a second venture into the use of design consultants, with John Claridge, many years later.

But it was the home front that really mattered.

Chapter Ten

New Team: Home Trade 1956/1959

Though in the four and a half years to the end of 1959 the export trade, until quotas and labelling spoiled the United States market, did relatively well, the home trade was the cause of constant concern.

Gardiner's policy was, and had ever been, to sell through agencies, the agent's efforts being backed up by frequent visits to markets and customers by the Director in charge of design and sales, in this case Colin Brown. The firm which represented Gardiner in London, Alan Shepherd Ltd., was old established, substantial and one of the top flight of London agencies. Founded nearly half a century earlier by Alan Shepherd it was now owned and managed by two "partners", Ken Smith and Gerald Box. Ken had joined before the war, had gone off to do his bit in the R.N.V.R. returning on demobilisation as Lt. Commander Smith, textile salesman. Gerald Box, too young for war service, had done his National Service in the Indian Army and for many years continued on as reservist Major Box in the Artists Rifles. For a while, before being invited to join Shepherd's, he worked at the famous merchant house of Jacqmar. In those days the agency had, besides these two active partners, one salesman, a secretary/accountant, Kay Watson, and a typist. They occupied the fifth, top, floor of 17 Great Portland Street, the heart of the wool textile business in London, the surrounding streets being occupied by the London offices of the major garment making firms. Mornessa, Hebe Sports, London Maid, Crayson, Koupy, Harella, Michael Nadler and the like were within a stone's throw. Others such as Steinberg, Dereta and Ellis and Goldstein were further afield, in the East End.

Besides Gardiner, Shepherd represented John Knox of Silsden, a Yorkshire worsted mill, and Lesur, a very high novelty French firm.

In the provinces Gardiners were represented by Walter Sinclair, formerly a Director of Gardiner, and his son Roddy. The Walter

78

Sinclair business had been set up in offices in Regent Street when Walter left Selkirk and represented two Yorkshire mills in London, and Gardiner in the provinces. No doubt the London agency would have suited them better, but it was spoken for.

Both agents sold women's wear and men's wear fabrics, but as men's wear styling was all North American oriented it was the former which was the basis of operations.

Home trade operations in the woollen, as distinct from the worsted, field, where Scotland occupied almost unchallenged the field of highly styled twist worsted fabrics, were conducted against the background of Yorkshire dominance. The mills there were substantially larger than their Scottish counterparts and generally had two strings to their bow, pure new wool cloths, and at a lower price level fabrics made from recycled wool and other fibres whose Victorian names, from which Yorkshire successfully escaped, were Shoddy and Mungo. The wool fibre is remarkably tough and the principal degradation in it's recycling is that the processes break the fibres into shorter lengths, which poses spinning problems which Yorkshiremen were adept at solving. It was said of them that "if it had two ends it could be spun"! Today's Greens must be in a dilemma. Recycling is good because it preserves the world's resources. But recycling is bad because, by making animals redundant it disturbs the balance of nature. Perhaps so.

At this time synthetics were in their infancy. Nylon was well established, Rayon had been about since the thirties and Terylene, not yet generically called polyester, had appeared on the scene, and Yorkshire was far more willing to use them than was Scotland. Typical of this flexibility was their approach to the "Mohair loop" group of fabrics. These loop yarns were part of a strong fashion in the fifties in ladies coatings and suitings. "Properly", they were made from Mohair fibre in the loop, and wool worsted in the core and binder threads which held the loop together as a loop. This made for an expensive yarn. The cost effective Yorkshire answer was a wool and mohair, or even just wool, loop, with either a wool core and nylon binder, or both components in nylon. Either way the resultant yarn was a lot cheaper than the "proper" article and, though different in handle was, visually without a magnifying glass, indistinguishable from the real thing.

Quite apart from these differences in the conduct of Yorkshire and Scottish operations economies of scale made the southern fabrics, like for like, less expensive. Scotland, therefore, had to outdesign the Tykes and, by reason of the economics of the pricing of fashion garments, were selling into a more limited market, which meant smaller production runs which made for less efficient processing. And if a Scottish design was truly attractive to a larger scale garment maker there was always the possibility that he would have it copied in a cheaper fabric the originator getting perhaps, for his pains, just the sample order. This was a real danger, those makers prone to the practice were known and were, of course, avoided, limiting still further the range of potential customers.

When "the lads" took up where McNairn left off the basis of the business was an expanding upmarket North American men's wear trade, a limited European mainly women's wear trade, and a home trade with three legs, all women's wear. Firstly, the standard but difficult Crepe Tweed cloth was still running well, though the writing was on the wall for designs were involving more and more opposite twists of different colours; the over complication of design is often the harbinger of the demise of a fabric. Allied to the Crepes was a standard Chenille, or small loop fabric, which in terms of complication was going the same way. Then there was the pile fabric range, which was really in direct competition with Yorkshire but which, using as it did Mohair and Angora as components, had a label attraction and was still running. The third leg was the novelty fabric range, where loops, and slub yarns, and Donegals and other yarns with spots and irregularities were used with straight yarns to create the effects. These fabrics, less easily copied, carried higher margins than standard cloths but had built in penalties in the form of design costs, yarn stock carrying and obsolescence problems and the not inconsiderable risk of guessing wrong for any particular season.

The first Board meeting after McNairn's departure set the tone, with an extended discussion of the balance between home and export trade, and recognition of the fact that though the latter was extremely profitable it was vulnerable, so the home trade must be developed. Philip Robinson, rightly, underlined the need to extend men's wear operations from America into the home trade but failed to recognise that different cloths would be required. In one of his many interventions into day to day operations he said he would arrange introductions for Colin to the heads of Sumrie, (Monty Weir), and Chester Barrie, (Myron Ackermann), whom he knew. So did Colin, but PR in full flight was not to be denied. He also arranged a meeting with the manager of Lewis's factory, which made boys clothing and which might use cloth which Gardiner could make. As the year was promising well these were all constructive proposals, not defensive thoughts, though in the end nothing worth while came of any of them.

At the same meeting the idea of reducing the total number of customers so as to give the remainder better service was mooted. Here again Philip put his oar in, saying that he had seen certain customers who were capable of quantity buying but whose accounts with Gardiner had languished but who he hoped to bring back into the fold. A review of the customer list was proposed, but no active steps followed. These were the first of a long series of interventions by Philip Robinson, all well intentioned, but all made against the background of retail experience rather than wool textile/ manufacturing knowledge. It would be neither unfair or unkind to say that in these early years of the Brown Jackson regime Philip regarded himself as and acted rather like a part time Managing Director, perhaps with a view to nursing us through the years during which we learned Gardiner's trade. Tough, for he didn't know it either. Nor did any of his colleagues, for the McNairn years had been mainly volatile.

At about this time the NASWM hit upon the idea of exploiting the unique aspects of the Scottish product by means of a joint advertising campaign based on "woven in Scotland" and "pure new wool", and funded by the members with matching funding from the International Wool Secretariat, the body set up by wool producers to promote their fibre. I well remember the meeting called to discuss the proposal at the North British Hotel in Edinburgh. In those days the Industry was large enough to require the ballroom or a large meeting room for such an occasion, and I was there because Colin, whose province such things were, was abroad. Neither of us was enthusiastic, for we couldn't see how the raw material for another industry could be effectively promoted to that other industry's customers. So when, after a long discussion during which it became plain that the Industry would proceed, the vote came I qualified the Gardiner vote, saying that whilst probably in favour (I thought we couldn't be left out) I would have to consult my colleagues before going firm. At that time Colin and I were very new and regarded as young whippersnappers by the establishment, notwithstanding Colin's Border lineage, and I earned a rebuke from the Chairman for my trouble. James Scott-Noble took me to task, saying that he had assumed that everyone present was empowered to act for his Company. I duly squirmed. But when it came to the bit Gardiner did join, not from conviction but simply not to be the odd man out, and went on to make more use of what became the Scottish Woollen Publicity Council, and get more benefit from it than any other member. So much so that there were from time to time mutterings about our hogging the action!

In August of 1955 Philip Robinson mounted the first of many attacks on our home trade agents, noting that to make the year a real success we needed £30,000 of sales over and above those in prospect, going on to say that he couldn't see the agents getting them, so perhaps the Mill should appoint it's own traveller to fill the gap. In the end the agreement was that Colin and Philip should visit Shepherds and Sinclairs, give them a severe wigging and unveil the threat of appointing a traveller. The mechanics of such a mixed selling arrangement were left in the air. At the same meeting the urgent installation of four of the twelve looms on order was agreed, as was the appointment of Gardiner's first time study man, Peter Bond.

The overdraft limit at this time was £65,000, but the expanding cashmere business had it pushing the limit too much of the time so Hugh Watson set up a meeting with David Alexander, then General Manager of the National Bank of Scotland, and Colin and me to discuss a temporary limit of £80,000. This was agreed, and that meeting was the first of an almost unbroken series of meetings over thirty one years at which the year's accounts were discussed, the facility for the ensuing year was agreed and the Bank used us as sounding board into the Industry. Always, bar once, followed by lunch in the Bank dining room, these outings were valuable and enjoyable. They were usually in May, and Colin and I fell into the way of taking our golf clubs and having a game at Gullane afterwards, "it being hardly worth going back to the mill". In later years the Bank found out about this, and several times the General Manager, accompanied by a suitably good colleague, came down to the coast for a game with us. Perhaps it did something to keep relations with the much maligned Banking fraternity on an even keel through thick and thin.

In October the customer reduction policy was replaced by instructions to the agents to go out beyond their normal clientele to try to sell small quantities on a scattergun principle to fill the looming gap. This time Philip's idea was not to engage a traveller, but to see whether there was anyone in the mill who could go out on the road and get a few orders. There wasn't. The meeting did agree to appoint an agent, Stockland and Ferguson, for Hong Kong, Japan, Syria, Lebanon, Egypt and Iraq. For a few years they did remarkably well, principally in Japan.

Despite the niggles about agents the year was going well, albeit the future looked less rosy. Twenty more spindles of Supercop winding, an ingiving machine, a Burroughs Sensimatic accounting machine and what is optimistically described as a photocopying machine were authorised. This was in fact an Ozalid diazo copier, a messy machine producing wet copies of mixed quality. But it did mark the first steps away from the hand writing of "piece tickets", the document which accompanied each group of pieces warped together as far as the weaving shed, which contained all the details needed to weave the beam. A short step on the long road to computer generation of all the production documents.

The last meeting of the financial year to January 1956 brought more strife for the agents. Second half orders were below the previous year's and Philip Robinson took on the job of seeing the agents again and reading the riot act in general, and in particular demanding £25,000 of extra sales from Shepherd by the end of July, and £15,000 from Sinclair. London representation was discussed in general and though no immediate change was proposed the idea that it could not go on like this indefinitely was left hanging in the air.

Payment of the Preference dividend was agreed, and at the next meeting, at which the accounts showing a profit of £30,810, were considered, a transfer of £4,000 to the Capital Redemption Reserve Fund, which would allow the arrears of Preference redemption to be made up, was agreed.

But before the accounts were discussed the home trade came in for another attack. Whilst the Board stuck to their decision to let the agencies stand for another year the traveller idea was promoted with vigour, with Philip offering to use his contacts to find a suitable person, failing which the post would be advertised. This time the agents were to be told, and the allocation of accounts and customers worked out with them. As ever, it all came to naught, but the discussion went on to detail action to be taken at mill level.

Men's wear cloths were to be developed suitable for use by the larger garment makers, and whatever new yarns were needed were to be provided. This project had a long gestation period and took some fifteen years to reach maturity. Then it was agreed that Gardiner should try to resume dealings with Marks & Spencer, discontinued since Utility days. Surprisingly Sir James Campbell had the entree here, and was to be provided with suitable samples. Much discussion ended in agreement that Marks & Spencer should be quoted cost prices, and that agents should be told they would get no commission on these dealings. This decision is, perhaps, the first tiny step in the direction in which the Company was to go, with success, in the sixties.

Thirdly, Philip Robinson returned to the charge with the suggestion that there was a lot of business to be done in the childrens and teenage field, with an offer to introduce Colin to these large operators, or those of them that he knew. A couple of months later Philip, at a meeting where yet again it had to be noted that while export sales were up, home trade were down, proposed that he set up a meeting for Colin with the buyers at C & A. An entree there, if effective, would have been highly desirable but as with so many of the initiatives of the last year the idea was flawed, failing as it did to recognise that with less than sixty looms in commission and with Gardiner's diverse product range the cost structure was inappropriate to the conduct of profitable business with a firm like C & A, or with Marks & Spencer for that matter. The laudable objective was to recover otherwise unrecovered costs, but on their own neither of these proposals could contribute to profit growth.

The last four of the twelve surprise looms were ordered for September 1956 delivery, which would take the capacity beyond sixty again, and continued reasonable prosperity allowed the ordering to the new Tenter. Ideas of using discounted Bills of Exchange to bypass the overdraft were floated, and discarded on learning that the Bank would call them overdraft if they discounted them and would discourage us from discounting them elsewhere. Preference dividends were maintained and, on 24 July 1956 William Miller Calderhead, Managing Director of UTR Ltd., Henry Colin Brown and Ian McKenzie Jackson were elected to the Board.

Calderhead's appointment was because of his textile background and because of his active involvement in projects such as the abortive transfer of finishing and the Lennox joint venture. The appointment unbalanced the Board, for UTR now had two representatives to Selfridge's one. It may be that this should have been seen as the first indication of Selfridge's loss of interest in Gardiner.

The first meeting of the new Board was on 27 August 1956, by which time the half year's accounts should have been ready, but they were not, for a severe epidemic of summer 'flu had made stocktaking impossible. But the standard costing/accounting system was working properly by now so the probable result was known within a narrow band. There was comment on the size of the adverse volume variance and on it's persistence, as well as the usual home and export comparison. Yet another reading of the riot act to the agents was proposed for 12 September in Philip Robinson's office with PR, Calderhead and Colin to meet two persons from each agency. And there was one other subject to be put to this sales meeting.

Against the background of the persistent adverse volume variance the Management proposed that all home trade prices except for those of a few established successful cloths to be reduced by 4%, the arithmetic showing that an increase in volume believed to be attainable as a result would more than make up in overhead recovery for the lost margin. The sole objector was Philip Robinson, who regarded it as too much of a gamble, and believed that our selling arrangements were our weakness and that if they could be made properly effective the desired volume could be achieved without sacrificing margin. He did, however, agree that a scale of reductions for bulk orders be devised, and put to the agents in September.

Lennox reared it's head again with Calderhead asking whether, should their American distributor like to have matching skirts made from Gardiner cloth, a conflict with George Klein would arise. The answer being almost certainly yes this one was left in the air. The idea surfaced again at the September meeting, when it was agreed that Colin and Calderhead, who would shortly be in New York at the same time, explore the idea with George Klein. Thereafter the idea disappeared from the scene.

The price reduction scheme settled to a 4% drop at the agents's discretion where 18 pieces to a range were ordered, and where the agent felt that the offer would result in orders totalling 50 pieces to the range. Crepes and Chenilles were excluded.

The rest of that year was dominated by the cashmere labelling and quota problems in the USA and by the loss of Greiff as a customer, though each minute contains some reference to the decline of home trade sales and the performance failure of the agents. Sinclair was excluded from the development of new men's wear cloths, and their sale effectively taken out of his hands and made the mill's province.

By the end of 1957 the management had been working together without a Managing Director for a year and three quarters, and working for Gardiner for three years and two and a quarter years (Colin) respectively. The first nine months of their tandem operation had seen the completion of a successful year whose foundations had been laid by McNairn. The rest of the period had produced a less successful year beset by problems in the lucrative American market and bedevilled by falling home trade orders. In textiles you learn quickly, or not at all, and by the time of the January Board meeting in 1957 the management had begun to form a background of belief about the position of Gardiner in the industry and but yet only vaguely formulated ideas as to future direction. While this process was going on the shareholders, Selfridges in particular, were beginning to lose interest, unable to see where a profitable future lay. Philip Robinson missed the January meeting, and his absences became more frequent as time went on. His non appearance in January was most unfortunate.

At that meeting the management put forward two propositions. Firstly, that sacking agents would be counter productive, and secondly, that to sell in bulk to the larger garment makers would require both the making of cloths specially for the purpose and, the cost structure being what it was, selling those cloths at cost and with a reduced commission for the agent. Management recommended that this proposal be implemented and cited the loss of Greiff and the looming quota problems as supportive of urgent action. The shortfall of pieces with those problems looked to be of the order of 2500, or nearly half the target capacity of the mill.

In the absence of Philip Robinson the Board felt unable to come to a conclusion. The possibility of such a policy starting a price war was floated and Calderhead asked whether such an operation would do anything to mitigate the seasonality which was bedevilling the Company's trade. Sports jacketings, men's suitings, cloths for women's winter suits and skirts, along with novelty coatings, filled the first, or winter season half, but spring cloths to fill the second half were more difficult. The answer was "no", which may have helped the Board to defer a decision.

At the end of that month, with Philip Robinson present, the idea was discussed again and rejected, and a complicated substitute was put forward and adopted. Management were told to pay serious attention to the production of women's wear cloths which could be sold in bulk at normal margins, and should also devise men's wear cloths for the same purpose. Would that we could! As a stopgap Sinclair was to be told to sell 500 pieces of a particular coat/suit cloth at cost for half commission to large makers in the provinces. Proper nettle grasping was still a year or two away.

For all that every firm in the Scottish industry was the bitter competitor of all it's fellows, and the rival of it's Yorkshire counterparts, there was a lot of inter firm co-operation, and at this meeting there was discussed, and rejected, a proposal that Gardiner, Gibson and Lumgair, and Ettrick Mill should set up a joint central boilerhouse to provide steam for all three.

And the Director's fee fund was increased from £1000 to £1500. Sir James thought that "the labourer is worthy of his hire".

The year ended with a profit of £17,922.

The following year, 1957/58, was a year of declining sales and inadequate or ineffective attempts to restore the position. By July Sinclair had used 278 of his cost price pieces, and by January 1958 - 505, when another 500 were authorised. But relations with him were less than ideal. Though holding Gardiner's agency for the provinces his office was in Regent Street, and he had no office in the primary provincial garment making centre of Leeds. In July it was reported that a number of customers had said that they didn't want to be called on by Walter Sinclair, but that they had no objection to his son Roderick. With this information, and with the need for a Leeds office deemed paramount, the Board agreed to tell Sinclairs that Roddy had to move to Leeds and open an office there, or lose the agency. Philip Robinson was not present at this meeting.

The idea was immediately rejected out of hand by the Sinclairs but, typically of the Board at the time, immediate steps to find another agent were not taken and the unsatisfactory situation dragged on.

There was good news in October, when there was an improvement in Alan Shepherd's orders, but the Gardiner Board were still scratching around for ideas, and Philip Robinson's suggestion that a firm with which he was connected, Portman Wholesale, which bought for a group of eight continental department store chains, might sell our product on the Continent, was taken seriously. Like all such ideas nothing came of it.

Light relief during the year was brought by two incidents, both of which underlined Selfridges' loss of interest. At Philip Robinson's request Colin was asked to show an unidentified person round the mill, for an undisclosed purpose. Heavily disguised in a homburg hat, dark overcoat and sunglasses he was identified as David Clore, whose brother Charles was then at the peak of his take over career. The other was a visit to the mill by Sydney Leake, Chairman of Selfridges. The visit itself was without incident or profit, but Hugh Watson quietly told us when he next saw us that Leake had said to him in the car on the way down from Edinburgh: - "Watson, what am I, Chairman of a group of retail stores, doing in Scotland visiting a tweed mill which makes nothing for us and pays us no dividends?"

At the end of the year, with things looking rough, there was a long talk about the desirability of maintaining a plant with a capacity larger than appeared to be saleable, and of the alternative of chopping it in size and budgeting accordingly. Fortunately, by this time, as a direct result of my having agreed to speak about the financial aspects of woollen manufacturing to the Former Students Guild of the Scottish Woollen Technical College, I had formulated the theory that there was no place for a mill with between 20 and 60 looms. Below that figure short run exclusive fabrics could be profitably made, as the experience of Ebenezer Johnston demonstrated. Above that, the medium scale garment makers could be sold to. In between was a desert, with costs too high for working with the medium makers, and too high a capacity for the short run trade. All this leaves out entirely the problems of manning for more than one type of trade. In the end maintenance of capacity with short time working if needed was the preferred option.

The year to 31 January 1958 ended with a profit of £2,598. The Preference dividends and redemptions were up to date.

The following year, 1958/59, was even more difficult, with recession compounding the effects of Gardiner's fundamental problems of size and market positioning. By the August Board meeting, by which time it was known that a small loss had been incurred in the first half the Chairman was prompted to ask, in so many words, "what are the long term prospects for the mill?" The management responded with a proposal which, like so many before it, held out alluring prospects but eventually came to nothing.

At that time there was a garment maker conducting a very successful promotion of coats made from a velour fabric and sold under the name "Kashmoor". Whether the cloth had any cashmere content we never established, but the promotion clearly worked. The management proposal was that using an established Gardiner wool and angora velour cloth an approach be made to a large garment maker with a proposal that he and Gardiner run a similar operation on a joint basis sharing the promotion costs. We recognised that the price would have to be cut to the bone, but real volume on a simple cloth would do wonders for the volume variance.

Even Philip Robinson was enthusiastic, despite the sacrifice of margin, and offered to have sample garments made and photographed so that we should have more than just bits of cloth and an idea with which to approach the maker. From a list of possible partners which included Windsmoor and Mornessa, we selected Steinberg, selling under the name "Alexon", for the first approach.

We were received with interest. The concept appealed, but in the end the best price at which we could offer the cloth was deemed by Steinberg to be too high to end up in a mass marketable garment, and the idea foundered. Other approaches were made with the same result, though at the December meeting Philip Robinson said that Mr. Seigal had expressed interest in a limited operation and Colin agreed to meet him when next he was in London. Nothing came of that; Seigal was a merchant in a very big way operating at price levels which made even Yorkshire mills stay away if possible, though the volumes he generated were very substantial.

At that meeting some small expenditure of capital was agreed. A conveyor to move pieces up to the first floor darning flat, some beam racking and some Assembly winding machinery, to serve the twisting operation and with the potential to increase it's output quite dramatically.

The year closed with an improved second half and a loss of only £88. It is not too surprising that at the March 1959 meeting at which the year's accounts were reviewed the management's proposal that ten old and decrepit looms be scrapped and ten new bought was not accepted. Four new were agreed, with Philip Robinson dissenting.

1959 began better that 1958. The Far East began to produce useful orders, Roosen was prospering in Holland and the Forstmann operation was working, though not without it's troubles. One of their cloths in particular, containing very brittle Reindeer hair, played havoc with weaving efficiency.

By September the Board's patience with Sinclair was exhausted. His correct statement at one of the numerous discussions that "the product was not right" did not get the attention it deserved, and did his cause no good. After researching and interviewing a man with a view to his joining Gardiner as provincial traveller, and turning him down, the agency firm of Chilton, Phillis and Guest, of Huddersfield, were appointed in November to replace Sinclair.

Throughout the year the management had been a trifle unsettled. The disinterest of the shareholders, stemming directly from lack of profits, dividends and prospects, was becoming plainer by the month. Despite the lack of progress and the difficulties encountered the prospect of the owners deciding that they had had enough and closing the mill before all was lost, one which we deemed very real, was not attractive. Lurking in our bosoms was the feeling that there was an answer, and that we were close to finding it. So, after reviewing our personal financial situations Colin and I approached both Selfridges and UTR with a proposal that we buy their holdings, splitting them equally between us. For some reason two separate sets of negotiations were carried through, those with Selfridges resulting in an offer of eight shillings a share, which we agreed to. Those with UTR stuck at thirty shillings a share, which we could not agree to. So, on 23 November 1959 the Selfridge holding was transferred half to Colin and half to me, with UTR retaining the other fifty percent, a shareholding pattern which was to persist for twenty seven years.

Philip Robinson resigned from the Board. We entered into a shareholding agreement with UTR which gave each party a pre-emptive option should a sale to an outsider be contemplated. The agreement was written by our independent Chairman, Sir Hugh Watson and, providing as it did for our purchase rights to be two separate options, was heavily biased in favour of "the lads". Colin was summoned to Glasgow by Hugh Cowan-Douglas to discuss it, and, after a heavy and liquid lunch in the Great Western Club it was signed, unread by HCD. He died two weeks later. Perhaps he had read it!

Now we really had to find the answers.

Chapter Eleven

Finding the Answer 1959/1961

The year of our financial involvement, ending on 31 January 1960, yielded a profit of £14,690, a result which we later came to regard as having delayed our coming to the conclusions which set Gardiner on the road to prosperity. The result was achieved on the back of a United States market which had not yet failed, the Forstmann operation and a marginally improved home trade. This latter owed it's relative buoyancy to the then prominent merchanting house of Jacqmar, whose orders with Gardiner were the best ever. Though Jacqmar today is but a remnant of the once great firm spun off from failure it was then big and strong enough to contemplate buying the United Turkey Red Company, then fighting off the prospect of liquidation. The negotiations came to naught and early in 1960 UTR was bought by the Calico Printers Association of Manchester (CPA).

Just as UTR had once been great in the Scottish textile scene, so had CPA been great in the Lancashire picture. A gather up of merchant converters and cotton fabric printers they owned, in 1960, the largest printing works in Europe, as well as weaving and spinning plants throughout Lancashire. Their purchase of UTR was made with a view to shutting down most of it and indeed it did not take long for that once great Company to be reduced to just one works in Newton Mearns. CPA's traditional textile operations were as UTR's had been, struggling, their relative financial strength stemming entirely from royalties pouring in from all over the world on the patents of "Terylene", which had been invented or discovered by a scientist in their laboratories and without which they would quite possibly have gone the same way as UTR. The man who brought this life sustaining benefit to them was ill rewarded, for, as Colin and I were to discover, CPA's views on management remuneration were parsimonious in the extreme.

As the new owners of UTR scratched through the books and turned over stones to find what they had bought they came across the fifty percent shareholding in Edward Gardiner and Sons Ltd., standing in at £37,000. In August 1960 a senior Manchester Director, Mr. N. E. W. Hutchinson, attended a Gardiner Board meeting with a view to finding out just what they had found under that particular stone. He found falling sales figures and heard the Board approve the purchase of a triple headed Sellers cropping machine and associated flock collecting plant, for £3,000. Sir James Campbell, having lost all his UTR offices in the takeover, resigned from the Gardiner Board and in November Geoffry Collier Potter, Main Board Director, joined us for a stint which was to last eleven and a half years, ending when restructuring of yet another merger brought about his early retirement.

After these manoeuvres the ownership of Gardiner was as follows: -

The Calico Printers Association Ltd.	40,000	Ordinaries	50%
Henry Colin Brown	20,000	"	25%
Ian McKenzie Jackson	20,000	"	25%
15 Preference Shareholders	32,000	Preference Shares	

And, with the shareholding agreement requiring that there be an independent Chairman, the composition of the Board was: -

Sir Hugh Watson D.K.S. Chairman
W. M. Calderhead of UTR
S. C. Potter of CPA
Colin Brown
Ian Jackson

CPA were entitled to one more member, but after Bill Calderhead's resignation did not exercise that right until January 1965, when
R. T. Earnshaw was appointed.

As the year 1960, whose end would be 31 January 1961, rolled on it became clear that the previous year's result was not going to be matched. The first half lost £8817 against a profit of £9684 and first half losses boded ill, for the seasonal nature of the trade meant that lack of a first half profit almost certainly meant a whole year loss. The cause of the loss was the usual one, working below capacity despite the modest success of the reduced price schemes and the achievement by the new provincial agents of a substantial sports jacketing order from a new midlands customer, Foster Brothers, a good order which was to prove to be a "one off". With our own money now at risk Colin and I redoubled our efforts to devise a way of making profits on a consistent basis. Conferences with all our agents brought the depressing message that, in their opinion, price reductions on the present product range would not bring the volume required. The standard bread and butter cloths were running out of steam. Crepe Tweed and the velours were falling victim to the price effectiveness of the Yorkshire mills and the Chenilles were fading out of fashion, which left the novelty fabrics and American sports jacketings, both areas requiring enormous investment in design and pattern making and disproportionate investment in yarn stock.

The women's wear sub agent which George Klein had been persuaded to allow us had proved a failure and in November the tariff and quota arrangements in the United States were replaced by a simple duty of 38% plus 37½ cents per pound of weight, a formidable barrier. It was becoming obvious that unless something rather nice and unexpected happened the Preference redemption scheduled for 30 April 1961 would fail for lack of distributable reserves. As so often in textiles, things were looking bleak.

Though we were effectively partners our different backgrounds and experience made the relationship between me and Colin an uneasy one at this time, an unease which the stress generated by the looming problems did little to alleviate. At that time we had no structure of formal meetings to discuss the problems; formal consideration usually took place at Board meetings, or when there was a third party, such as an agent, present, whose comments and questions forced things into the open. But the real forum for airing the problems and ideas for their solution was a completely informal daily meeting between us in my office, when we sat on opposite sides of the desk and opened the mail together, something we did for the whole of our tenure. The mail itself provoked discussion and threw up ideas for pursuit which led to the generation of ideas of our own.

Our agents had given us a clear background against which to develop a prosperity plan. Price would not sell the current product in volume. Sinclair had said that the product was wrong anyway. The larger Yorkshire mills were doing better than we were. Exports, particularly the USA where we had achieved our best successes, were getting more and more difficult. The developing outcome of fiscal 61 was looming more disastrous by the week and nothing currently in prospect looked likely to improve the succeeding year.

Bill Calderhead left UTR and resigned from the Gardiner Board in November.

Opening the mail took longer and longer as Colin and I struggled with the problem. Each week saw another set of "what if" figures projecting the probable result of alternative courses of action. Too many of them forecast continuing losses, but through the chat and the figures a picture was beginning to emerge. The conclusion which we reached was far removed from the ideas of today's "market driven" philosophers, but yet, strangely and in a perverse way, close to them. We tackled the problem backwards, in their terms. Our question was, "what can we make that will sell in profitable volume"? Theirs would be "what can we sell in profitable volume and how shall we make it?" I suppose that, at the end of the day, the two questions are the same.

In the end the battle plan was extremely simple but not without risk. The background of potential customers and the Yorkshire competition said that price was, despite our agent's advice, crucial to the solution. Price and volume are indivisible bits of the commercial equation and our search was for the product which we could make on the existing plant in sufficient quantity to allow us to get within the price brackets of the middle volume garment makers and still allow us to be profitable.

At that time we were producing about a hundred pieces (of 60/65 yards) each week. Included in the product mix were cloths of coating, womens suiting and skirting weights made from our basic twofold sixteen cut cheviot yarn, and others made from the other basic yarn, twofold 24 cut saxony. The picks per inch of these cloths ranged from 12 to 20, with the average towards the lower end of the bracket. As basic cloths they were used in a limited way by a number of garment makers, and they were the cloths which had been bought, albeit in minimal quantity, by Marks & Spencer.

The "what if" equations showed that making nothing but these cloths, and limiting the minimum warp which we would accept to six pieces, or 420 yards warped, our kit of 62 operational looms could be made to produce 300 pieces per week working single shift only. When this volume of output was related to manning and expense budgets framed to match it the price field which was generated was 25% less than we were then quoting for the same fabrics, after building in a respectable margin for Gardiner. This, on the basic cloth, known then as SS 145 (standard skirting, 14.5 ounces per yard), later re-christened CP 31 (cheviot plaincloth, 310 grammes per square metre), amounted to four shillings a yard, an impressive reduction.

Output at that level, though within the capacity of the weaving shed and most of the pre-weaving processes, was beyond the reach of the finishing departments, and of the yarn twisting operation, so some investment in plant would be needed. But the more we sat across the desk in the mornings mulling over the options the more we realised that "up to yet", nothing else proposed or thought of had any potential to save our skins. The indicated price level would put us in direct confrontation with Yorkshire, and embarrass our Scottish colleagues to no small degree. And initial losses until the volume built up were inevitable, as were losses on all the redundant yarn stocks which would be created. Should the target volume not be achieved by anything more than a small percentage closing time would come upon us very quickly.

So we explored the idea of a mixed operation, continuing the existing business and going for volume through price on the selected cloths, but the arithmetic showed that the odd cloths would hold back overall production to such a degree that real volume on the specials would not be achievable, so the price reductions would not be there to be offered, so the volume targets would become even more unattainable. In our view, it was muck or nettles.

The accounts of the year to 31 January 1961 showed a loss of £20, 212. The meeting on 3 March at which they were discussed decided that with the loss just sustained, a Profit and Loss account balance of only £6508, and losses projected for the foreseeable future, the Preference redemption due on 30 April could not be made. And Colin and I presented to the Board for the first time our action plan.

After an extended discussion during which the only alternative suggestion, an export drive to fill the gaps, was discarded as being impractical with trade barriers in our prime markets at their present high levels, it was agreed that Potter and Brown should visit a CPA garment making subsidiary and test the water. Meanwhile, we would try it out on the agents who would have to do the volume selling. Decision day was set for the next Board meeting on 5 April.

During March the home trade agents were called to the mill and consulted. The cloths upon which the plan was based were unsuitable for export, so it was their opinion which would carry the weight. Chilton, Phillis and Guest were quietly optimistic, thinking that, properly styled, the fabrics could do well in the sports jacket trade. Their view of the women's wear aspect was less sanguine, for they held the view that it was rapidly migrating to London.

It was the conference with Ken Smith and Gerald Box that really set our eyes twinkling. As the plan was outlined to them, and the cloths pushed and pulled across the table they began attaching telephone number type volumes to a whole host of garment makers with whom we dealt not at all or in tiny volume. But it was very plain that they believed that, at last, we had a product which could be sold in quantity at the prices we proposed and that the loss of income flowing from proposed cut in their commission rate would soon be made good in cash. The sixty four dollar question, would Marks and Spencer be a real prospect, was answered with a solid and delighted yes. It was, after all the long series of meetings critical of Smith and Box who had, in fact, been up against it with the wrong product at the wrong price, an invigorating and momentous occasion.

We left for drinks and dinner at the George and Abbotsford Hotel in Melrose in an elated condition, unmindful of the fact that this was an exploratory chat, and that the Board had not yet approved the plan. The fact was that we all knew that there was no viable alternative, that this was something we could hang our hats on and that if we could make a go of it prosperity would be ours. Liar dice was a game in fashion at the time, and as we played it in the bar that night the plan had the code name "High Straight" attached to it. A name which it kept throughout it's life, one by which it came to be known throughout the industry.

97

On 5 April the Board, at full strength, met to decide Gardiner's fate. Brown and Potter reported on their visit to H. J. Wilson, a meeting from which the signals were confused but not discouraging. The agent's reactions were set out and after a long discussion the inevitable conclusion was arrived at:-

" After considering these reports, and after considerable discussion in which the following points were recognised:-

1. That there appeared to be no alternative policy which held much prospect for success.

2. That a complete change of outlook for both the mill and it's selling representatives was involved.

3. That to sell 15,000 pieces a year we would probably depend almost entirely on new customers in a price bracket at present foreign to us.

4. That the mill would become a mass production unit rather than a producer of novelties in short runs.

5. That to achieve the efficiencies required to enable the appropriate quotations to be made, the range of cloths produced would have to be very rigidly restricted, particularly in terms of picks per inch.

It was agreed that the change in policy which would result in the offering of a small range of cloths based on our 16 cut and 24 cut yarns and priced on the basis of highly efficient production should be made. It was agreed that for a season at least there would have to be some mixture of the old policy and the new in order that the stocks of yarn which would become redundant could be disposed of as largely as possible in cloth".

Thus was the scene set. Making it work, getting the customers to believe that we could do it, getting the plant into the right configuration, finding the right balance of the workforce, ensuring appropriate supplies of yarn and a host of other things were to be dealt with in the coming months.

Getting the customer to believe it was one of the major hurdles. Part of the scheme was our demand for a minimum order, after sampling, of six pieces per colour, a new and unwelcome idea to our customers and based on the philosophy that if we had multiple prices for varying quantities the garment maker would cost on the highest price and defeat the object. The first real test of our credibility came when, after sampling, Lou Ritter, head of Dereta, rang Colin and placed an order for ten items, four pieces per colour, which Colin refused. Loud and unfriendly noises preceded a curt goodbye. But when next day the order turned up in the post it was for six pieces to a colour.

But first, the Bank, the National Commercial Bank of Scotland. The General Manager was still David Alexander, with whom Sir Hugh set up a meeting at which Colin and I set out the plan, the risk of failure and the rewards for success should it come. David Alexander agreed to back us to the extent of continuing the overdraft limit at £80,000. We could not have asked for more, and we regarded this as a milestone in our ongoing good relationship with the Bank.

The loss making canteen was closed. A member of the accounting staff left and was not replaced, and accounts became quarterly instead of monthly. Trans British Textiles were given protective notice. The most critical period ever in Gardiner's history had begun.

Chapter Twelve

Making it Work 1961/1965

The background against which "High Straight" had to be made to work was very different to the conditions in the garment making industry, or the cloth manufacturing industry, which prevail today.

In 1961 the domestic garment industry and, for that matter it's overseas counterpart, included a substantial number of medium to large sized firms, selling under their own names, advertising their own products and each consuming hundreds of thousands of pieces of cloth every year. Many of them no longer exist, for a variety of reasons. Some of their owners simply closed the firm and retired on the proceeds. Some applied the accumulated profits to property, and left the garment industry. The majority of the departures were caused by plain failure, easily achieved in an industry subject not only to the normal commercial pressures in a free enterprise society but also to the lottery called fashion.

Whereas Gardiner's former clientele had been the medium and smaller makers, the target for the new policy had to be the larger to medium firms of which, in those days, there were considerable numbers. At the top of the tree were Marks and Spencer, not garment makers themselves, but the biggest buyers of cloth, which they farmed out to be made up for them. Then there were a host of fashion houses, household names in their day but many of them now either gone or of less importance. Such names as Windsmoor, Mornessa, Harella, Hebe Sports, Susan Small, Horrockses, Julius, Steinberg (Alexon), Gor-Ray, Skirtex, Emcar, Michael Nadler (Reldan), Dereta, Rosner, Ellis and Goldstein, Charles Kuperstein (Koupy), Jaeger, Aquascutum, Simpsons and many many more. Then there were the great London Merchants, Dick and Goldsmith, Burt and Willis, M.O.T. and Warwick Woollens, with the retailer John Lewis doing a massive piece goods trade. Wherever you looked there were customers to approach, for as well as all these and the other established firms there were a raft of up and comers elbowing their way onto the stage.

All of these customers and potential customers employed highly skilled cloth buyers who knew the cloth making industry inside out. If they required a particular cloth, they knew where to look for it, and had a more than approximate idea of what it should cost, and herein lay two of the great challenges to the new policy. Firstly Gardiner had to become in fact the prime producer of the fabrics on which the policy was based, making those cloths, quality for quality, at the best price in the market place, and backing the price with service in terms of delivery and consistent quality. And secondly those expert, hard nosed buyers had to be convinced of Gardiner's credibility. Achievement of the first objective was easier than achievement of the second.

The June Board meeting, the die being cast, was a deck clearing and forward looking occasion. The United States market as we knew it would disappear, and Trans British Textiles were put on notice, clearing the way for an agent sympathetic to the new approach. Whilst the long range overdraft forecast showed that we should remain within our limit the imponderables such as the payment performance of a new set of customers, the delivery performance of our spinners and the cloth take in behaviour of firms at the delivery end of the chain made it less reliable than hitherto. So it was agreed that, should it become necessary, the shareholders would lend the Company £10,000 in proportion to their shareholdings.

On the machinery front some relocation was proposed and orders were authorised for forty spindles of Arundel Coulthard two for one twisting machinery with appropriate assembly winding to go with it. This was a novel type of twisting machine where the input package was a large cheese with the threads to be twisted lying parallel on it, and the output package being a large cone which could be used in the warping process without re-winding. The machine's most innovative feature was the fact that two turns of twist were inserted for every revolution of the spindle rather than the normal one giving obvious output benefits. At the other end of the mill the inadequacy of the "blowing" or pressing machinery was recognised and a Sellers Blower of the latest type was ordered. To conserve cash all this machinery was to be bought on hire purchase, and so began a long relationship with Lloyds and Scottish Finance, the hire purchase wing of the National Commercial Bank.

A build up of stocks of the two yarns retained under the policy was authorised, and, at last, the threat to Alan Shepherd was lifted when it was agreed that they should implement the policy in London. Fred Carlin's contract was ended, CPA having a Paris office which, it was thought, could provide Carlin type services.

The half year's accounts to July, reviewed in September, showed a horrible, but not unexpected £13,604 loss, partly due to the implementation of the new policy. Indeed, but for the effects of that implementation the loss would have been less than the corresponding £8,817 of the previous year, so whatever distress might have been felt was more than wiped out by the optimism engendered by the buoyant forecasts of Gardiner's prospects under High Straight. Apart from the home trade, where the brightest prospects were, topcoatings for Canada, in the new standard cloths, looked hopeful and even TBT had been constrained to enter into a new, one year agreement which gave Gardiner much more freedom to sell to American customers visiting Britain.

Experience thus far showed that the vast bulk of sales would be of cloths made from twofold yarn, so another two for one twister was provisionally ordered for February delivery. Again to be bought on the never never.

By now the winter 1963 collection, the first devoted wholly to the new policy, was being shown to the new prospective customers by our agents and it is a measure of the dramatic effect of the idea on Gardiner's prices that, when showing to Marks and Spencer, the biggest and toughest of them all, Colin simply put the standard price list on the table, and it was accepted! This situation prevailed for a surprising number of years but didn't last forever. And, since the marketing of this collection was also the marketing of the new policy, word soon got round of the suicidal level of Gardiner's prices, and forecasts of our demise were thick on the ground. Occasions, such as meetings of the NASWM, of which Colin was by now a Council member, brought upon us from our colleagues in the industry a mixture of pity and anger, the former for the lunatic mistake we were assumed to have made, and the latter for the effect which our new prices were having on our competitors. As to the buyers, they too found it hard to believe that a Scottish mill could so quote and survive, and treated us with a degree of suspicion which made them tentative in their ordering in this our first season. Hardly surprising, for a buyer would be foolish to commit himself in depth to an uncertain source of supply.

By January 1962 the need for the extra twisting capacity had forced the confirmation of the provisional order and prompted a review of the wet finishing processes. This revealed a need for two combined milling and scouring machines and a further "whizzer", or more properly, hydro extractor. In the event the combines were bought second hand from Lumb, Walshaw and White, a firm of second hand machinery specialists who, over the years, were to sell us quite a lot of kit, and the whizzer was found in one of CPA's closed dyehouses. As the twisting department was to work double day shifts to keep up with demand it was fitted with off peak electric heating to avoid the need to run the boiler at night. And, finally in that round of capital spending a selvedge trimming machine, and barrows and cones and tubes and all the bits and pieces which clothe basic machinery were ordered.

Roosen, trading as Kennedy Millar, had been under performing and was under notice, but the minute notes that this seemed to have spurred him into action. Germany was sampling heavily, and in fact did nothing but that until years later we found the right agent. Customers in that market bought a sample length, supported by droves of postcard sized patterns and, season after season in our case, nothing else. Very discouraging, and very bad for sample making costs and the ratio of pieces to samples. Belgium and France, in the hands of a new agent proposed by CPA were not performing at all.

The January meeting agreed that the Preference dividend due on the 31 could not be paid.

At about this time the first of the mergers of Border spinning and weaving mills came about. The leading light was Douglas Hood, then Managing Director of Wilson and Glenny, in Hawick. His vision of the future was of groupings of mills rationalising their production facilities but marketing their products under their separate names. The other members of this first merger were Gibson and Lumgair in Selkirk, and William Brown of Wilderbank in Galashiels. Gardiners were invited to join but, fortunately, we had sufficient faith in our answer to the problems to decline. Later, D. Ballantyne of Peebles and Henry Ballantyne and Sons of Walkerburn joined, William Brown closed and Gibson and Lumgair withdrew to go their separate way. Of that first grouping there now remain D. Ballantyne; trading as Robert Noble and part of the Dawson Group, and Gibson and Lumgair, marketing scarves made on commission.

Mergermania took hold, and the first was soon followed by the second, when Stewart Roberts, of George Roberts, the biggest mill in Selkirk, and Hunter Thorburn, of Walter Thorburn in Peebles, prevailed on James Scott Noble of Robert Noble in Hawick to join with them in the grouping which became known as Roberts Thorburn and Noble. Why Nobles were persuaded to join was a mystery to us, for of the three mills, it was the only profitable one. That grouping did not prosper, Roberts, and Noble have long since closed and, after a fire at Peebles and the purchase of a spinning mill in Huddersfield the group traded, and still trades, as RTN Yarns. A complicated chain of events in the early seventies involving the failure of a major wool supplier forced Hunter Thorburn to sell it to the huge knitwear firm Nottingham Manufacturing for £1; RTN Yarns still trade, but now as part of Coats Viyella.

By May things were going in the right direction but not as fast as the Board would have liked. The first half was running up to a result which was to be a much reduced loss of £7,297 leaving the Profit and Loss account in the red to the tune of £15,717, no comfort to the dividendless and redemptionless Preference shareholders. Gardiner was not yet achieving a break-even performance and expense reductions and redundancies aimed at achieving that were worked out for immediate action. This was not as bad as it seemed, for the variety reduction would in all probability have brought about the job reductions anyway. The field at Eyemouth was put up for sale as was one van, not needed as the redundancies included the Lilliesleaf employees.

TBT released us for women's wear, Gunter Roosen became our agent in Holland, personally, not via Kennedy Millar.

The Bank insisted on a £40,000 floating charge on the assets. This had not been sought before because, though common practice in England, floating charges had not, hitherto, been legal in Scotland.

From May 1961 to 26 February 1962 there was only one Board meeting at which the only business seems to have been the approval of the half year's accounts, though since it was in Edinburgh with everyone present much more must have been discussed than that. Perhaps the long gap flows from the fact that we were all, figuratively speaking, holding our breath while the new policy grew to success or crashed to failure.

The year's accounts to 31 January 1963, showing a profit of £6,653 following a second half profit of £13, 932, the highest ever, were approved and showed clearly that High Straight was working and the order book statistics supported this, so for the first time there was an atmosphere of open optimism. We didn't know it then, but that second half year marked the turning of the corner and, but for a hiccup in 1964/65 Gardiner went steadily forward, the next setback not occurring until 1969/70. Just to show that the Board was still in touch with the real world a reduction of Capital was discussed, and rejected, lawyers fees being one of the good reasons for not doing it.

July 1963, with a first half profit of £9,860 in the making, nice, but not enough to allow payment of the Preference dividend, saw us back in the market for machinery. Four second hand looms at £250 each (installed), a reconditioned scouring machine, a 40 spindle coning machine and a weft straightener for the tenter with a total bill of £2,275. Modest, but the looms reflected the fact that we were pushing our capacity limits by now. And this meeting marked the beginning of a long running row with our Calico Printers Colleague, and later colleagues, and later still with the CPA Chairman. Though perfectly willing to increase the salaries of the Mill Manager and the designers, Geoffrey Potter absolutely refused to countenance increases for Brown and Jackson who, at this point had been on the same pay for four years, but for the profit share which was now being earned. Part of the problem was that CPA, themselves, were notoriously mean in the rewarding of their executives and Colin and I suspected that we were, in their minds, in some bracket, and that if we broke out all hell would be loosed in the CPA.

Sir Hugh was with us in our arguments but with the CPA representation being one short would not weigh in and force a vote. But he did, quietly, suggest that there might be benefit to us, and even tax advantage, if the Company were to buy our cars from us, and provide them to us, which, after a decent interval for researching the proposition, was what happened. Colin's Humber Super Snipe was bought for £1,610, and my Sunbeam Rapier for £1,250, and it was to be 1990 before either of us had to buy himself another car.

The financial year to 31 January 1964 rolled on with sales and production developing well. When the profit was struck at £19,587 it was off sales of £521,678, the highest year's total yet which, grossed up to old price levels to make comparisons fair would have been nearly £700,000 to be compared with the old firm's best of £402,000. Those sales had yielded a profit of £30,810, so in terms of reward there was still some way to go.

The year saw a marked increase in wool prices, reflected in the first increase in cloth prices since the policy change, accompanied by heart searchings and strivings for greater productive efficiency. The now unused fancy twisting machinery was sold and new coning machinery bought. The relative merits of coal fired chain grate stoking of the boiler or it's conversion to oil firing were debated, the latter winning. The conversion was carried out by Norman Sykes, the CPA Chief Engineer and the boiler lit one bitter winter's night, when the oil was so gluey that Norman thought it would never flow and light, but it did. Crepe Tweed not being a current product the Hattersley 375 looms were put up for sale, and the purchase of eight more standards debated, and in January an extra bay of yarn store building was authorised to accommodate the £20,000 of extra yarn stock required to keep production flowing. Another two for one twister, the third and last, was ordered and installed and eight more looms were fitted with automatic weft stop motions which, by "feeling" the pirn in the shuttle with brass probes, stopped the loom before the shuttle ran empty.

The year's profit allowed the arrears of Preference Dividend to be paid, but neither the redemption due in April 1964 nor of any of the arrears could be contemplated, for the Profit and Loss balance, after the dividends, was a mere £580. The by now obvious success of the new policies allowed the Bank to continue the £80,000 facility, and at that year's meeting with the general manager we were enjoined to "ask if we needed more".

But there was a minor recession lurking, and the winter 1964 season came to a grinding halt in July, leaving an unpluggable gap before spring 65 got under way, and it was this, and the less than buoyant spring season that caused the first half profit of £13,514 to drift away to a year's £9,298. But the Board was full of confidence and decided to expand operations in the men's wear field, employing an extra designer for the purpose, buying another combined milling and scouring machine and initiating male night shift weaving. The designer who joined for the menswear operation was Derek Garrett, a Hawick man who came to us from Yorkshire via South Africa, and ultimately left to manage one of our principal competitors, Laidlaw of Keith.

All this was clearly going to push the overdraft hard against it's limit, so the consent of the Preference Shareholders to an increased borrowing limit of £100,000 required by the Articles, was sought and obtained, and after that the Bank agreed to the new limit. Even so the Ordinary Shareholders agreed to stand by with £10,000 between them should it be required, the last proposed loans never having been called up.

And in July Geoffrey Potter refused to talk about the Executive Director's salaries. Again, perhaps as a consequence of this impasse CPA exercised their right to have two Directors by, in January 1965, appointing Reginald Earnshaw to the Board. He was not a CPA Main Board Director, but was the group's cost accounting wizard. He, too, had quaint ideas about executive salaries. It was at the time of his appointment that the arguments about consolidation with CPA's accounts, and the provision of schedules of figures to suit that purpose, began. To the day we sold the Company I refused to fill in their forms, and the proposition that we sell them one share, to give them more than fifty percent and make us consolidatable as a subsidiary will pop up more than once in this history.

Early in 1965, the model 375 looms having been sold, four second hand Hattersley Standards were bought, along with a new steam box for conditioning yarn after twisting.

The profit for the year to 31 January 1965 of £9,298 allowed not only the Preference Dividend to be paid, but also allowed the setting aside of £8,000 to clear off the arrears of redemption, but was inadequate to provide for the 1965 redemption. After these appropriations the P & L balance stood at £697, but we were slowly getting our feet out of the treacle. Not only were we restoring our reputations with our Preference Shareholders but our own industry was taking us seriously by now. Colin, already a member of the NASWM Council became Chairman of the Scottish Woollen Publicity Council in 1965, while I had become Chairman of the Employers Association in 1964. Could we say that "we had arrived?"

The high wool prices which had inhibited the year just ended fell in March 1965 and, though because of commercial uncertainty the season got off to a late start, the projections looked good. Our experience so far of the High Straight policy had confirmed our earlier view that it was a home trade operation, and that export would be limited and difficult. For the rest of the sixties our export percentage never topped five, and our agents had to struggle for that. When TBT's years agreement ended, Adam Watt, a Border emigrant to New York, had been appointed to represent Gardiner but, after only a short period in office, he fell upon hard times and had to part company with us. Though technically bankrupt Adam struggled on and eventually paid all his creditors, to his immense credit. His replacement was Joe Stiassni, who had masterminded the Forstmann operation with Gardiner.

Early in 1965 CPA bought two large London garment makers, London Maid and Alexander Newman, and Geoffrey Potter set out to do what he could to promote Gardiner fabrics with them. But, as ever, the managers of subsidiaries take unkindly to interference from the Main Board, and no particular benefits flowed our way. With our expanded output, and with the menswear operation beginning to show results, we ran into yarn supply problems. We had decided at the outset to concentrate our yarn buying on two spinners, Pogson of Slaithwaite, and Shires, of Milnsbridge, and as our needs grew they began to struggle. On this occasion one of the causes was the accidental concentration of the season's needs on one spinner, Pogson, who made the 16 cut yarn. All sorts of ideas were explored, joint stocking of yarn, out of season manufacture, sending spinning work out on commission and having both spinners make both qualities. For one reason or another no perfect solution was ever found, and in the course of frequent journeys to Yorkshire to "progress" matters (read the riot act), my friendship with Gerald Shires and Raymond Pogson in particular developed to a point which was to be embarrassing in 1974. Another cause of the 1965 supply problem was the concentration of the season on one colour, bleached white, which called for particularly white wool; which was always available, but when 25% of all our intake was in that colour supply did become tight. This high percentage of white usage lasted several years, so spinners and merchants learned to accommodate to it as time went by. Also at this time, after the shock of the previous year's increase in wool prices, we began to use futures to fix our prices. This was, theoretically, not practicable, for the primary futures contract was in merino wool, whereas we used crossbred. But, whilst they did not move exactly in step they generally moved in the same direction, so the use of futures for limited insurance did work. But not so well that we pursued it very far. One or other of the Israeli wars upset things so much that we gave it up.

The yarn supply problems prompted the Board, in April, to increase yarn stocks to the limit of the Company's resources, and to make it really work the shareholder's £10,000 of loans were called up.

The year to 31 January 1966 ended with a profit of £45,358, twelve thousand made in the first half and the rest in the second, one of the very few years in which the second half beat the first. When one considers the difficulty of filling the spring (second half) season from the seventies onwards this is worth explaining. The fact was that the High Straight product was more suited to womens suits and skirts than to winter coats, where novelty and heavy weight mattered more. The spring coat was another matter, the cloths being suitable for that product, which goes a long way to explaining the very heavy use of bleached white. We often joked that you could run a spring season on only four colours, red, navy, beige and bleached white!

So, the womans suit not yet having given way to denim rags, and it being a spring item just as much as a winter one, our spring seasons tended, for a while, to be easier than the winter ones. And from this flowed the concentration on bleached white.

This result allowed the redemption of the arrears of 2000 Preference shares, and of the 2000 due in April 1966. All arrears of dividend had been made good already, so the Board allowed the members the luxury of a 10% dividend on the Ordinary shares, the first for many a long year. If proof of the success of the High Straight policy were needed, there it was. Sales £685,515, Profit £45,358, Ordinary Dividend 10%. As always, the dividend was declared as an interim to avoid the need for the approval of a General Meeting, and in those days of the investment income surcharge to income tax it was waived by the executive directors, and added to their bonuses.

Though we had begun to prosper in a meaningful way it was on the back of a Victorian weaving plant, labour intensive in a world in which labour was getting more and more expensive. The machines to make weaving faster and cheaper, and better, were on the market. It was time to initiate the next gamble.

Chapter Thirteen

Re-Equipment 1966/1969

In 1960 the National Association of Scottish Woollen Manufacturers held it's first "Conference", at Gleneagles, the start of a bi-annual series which has continued to this day. Not always at Gleneagles, but always at one of the great Scottish Hotels, they are two day affairs assembling on Friday night, with seminars or lectures on Saturday, a Conference Dinner on Saturday night and optional golf on Sunday morning with dispersal before or after that, according to taste. Originally, while the industry remained comprised mainly of family businesses the participants were exclusively Directors and their wives, but in later years senior executives and designers also joined in the bi-annual "jolly", for that is what it is for all that is dressed up as work, with the AGM's of the associations lending an air of respectability.

For all that, the quality of some of the talks and the exchange of views with fellow strugglers in the industry made the Conference a useful forum as well as an opportunity to renew old acquaintanceships and friendships.

Most years the Conference had a theme, and at the 1966 affair, held in the Marine Hotel at Troon the theme was weaving, for in the nineteen fifties the first generation of modern weaving machinery had made it's appearance on the textile scene. Though the rapier loom, which was to dominate the seventies and eighties had yet to arrive the water jets, the air jets and, for the woollen industry, the Sulzer shuttleless weaving machine were all on the market. As always the machinery was primarily designed for and aimed at the much larger cotton industry, but the Sulzer machine had found a niche in the worsted industry, and was being explored in the woollen weaving field.

The principal difference between looms and weaving machines is that in a loom the weft is wound onto a pirn which is put into a shuttle which is batted back and forth until it is empty. The carrying capacity of a shuttle is relatively small, so stoppages to change the shuttle are frequent. Even before the war there were automatic looms on the market, the most prominent being the Northrop, which could, without pausing, sense the empty weft package and change it in the shuttle without pausing, but their working speed was no better than a conventional loom at about 120 picks per minute. This speed limitation was the result of the need to move the heavy race and reed backwards and forwards with every pick.

The Sulzer worked at some 240 picks per minute from a weft package of a kilogram or more placed at the side of the machine which could be "topped and tailed" with it's replacement. The end of the weft thread was gripped by a tweezer like gripper in a carrier about the size and shape of a pocket penknife, which was propelled across the face of the reed to the other side of the machine. On arrival, the leading end of the weft thread was tucked into the selvedge threads of the warp, while at the other end the weft thread was cut and tucked into the selvedge at that side of the machine, while the used carrier dropped onto a conveyor for return to the working side of the machine.

Not only was the speed of a Sulzer more than twice that of a conventional loom but it could work at a much higher efficiency because of the virtually endless weft supply and, for that very reason, whilst conventional weaving of the Gardiner product made two looms per weaver the maximum, eight Sulzers per weaver was deemed possible. The major limitations of the early Sulzers were the restriction to four weft colours, restrictions on overall pattern size and the need for strong thread, both warp and weft. And they cost £4,400 each against £600 for a Hattersley Standard six box loom which could, theoretically, weave up to 11 weft colours though in practical terms the limit was about eight.

Wilson and Glenny of Hawick, famed for their twist worsteds and thorn-proof woollens, were the first Scottish mill to buy Sulzers and by the date of the 1966 Conference they were installed and working, so who better to expound their virtues than Alex Goldie, Glenny's Production Director, and a most convincing case he made for them which was some comfort to me and Colin, for we had that very month placed an order for eight of them.

That decision was the logical extension of the bulk production policy which was now bearing fruit. 30 April 1966 was to see the Preference Redemption which would clear off the arrears and the outlook was good enough for the Executive to propose to the Board that the Sulzer step be taken, which we did at the meeting held at Sir Hugh's office in Edinburgh on 9 March. Sadly it was to be the last Gardiner Board meeting which he attended and it was one at which the usual convivial lunch was preceded by serious decision making.

The substantial paper under consideration set out the way we hoped to make the eight Sulzers work at 210 picks per minute and 77% efficiency. It went on to predict the outcome of various product mixes using the concept of "reward value". The R/V of a piece is it's value to the mill, that is to say the total value of the piece less the yarn content, agent's commission and anticipated cash discounts to be allowed. Using this concept it is a simple matter to value various product mixes and compare them with forecast expenses to get an idea of what is likely to work and what is likely to hurt.

We were conservative in our profit forecasts, £30,000 for the current year which, in the event yielded £63,824 and £24,000 for the following year which yielded £85,444. Small wonder that the members were not asked for more capital. How sad that Sir Hugh was not around to see those results.

Delivery of the eight Sulzers was to be April 1967 and there was a lot to be done before then. The prime question was where to put the building to house them. The Gardiner mill virtually filled the plot, so space had to found somewhere, and help came in the unlikely form of Doctor Beeching, who closed the Galashiels to Selkirk railway line, part of which was contiguous with and to the north of the Gardiner land. It's six mile length was put up for sale in chunks, one of which ran from the level crossing at Gibson and Lumgair's mill to the Gardiner western boundary with Yarrow mill. This we had to have for our weaving shed but we knew that Bernat Klein, then the owner of Gibson and Lumgair, who intended to buy the station for a craft centre or some such venture was after it as a link between it and

St. Mary's mill. And Bill Colledge, whose firm owned Yarrow mill to our west, and Ettrick mill on our eastern boundary, also wanted it, to run unlicensed lorries between his two properties.

It was a sale by tender, so there was little we could do bar put in the winning offer without going mad, and this we managed to do, so the new Gardiner weaving shed stands on the roadbed of the Selkirk to Galashiels railway line. This purchase was to stand us in good stead later, for this piece of railway line separated us from "Sim's" mill, to the north, owned by Turnbulls of Hawick, which we were later to buy. The railway land, had we not got it, would have split our property in two.

In April the Executive Directors were granted a salary increase, with no record of a fight. And this in Sir Hugh's absence.

By July we had planning permission for the shed, and the necessary Industrial Development Certificate, all the ancillary machinery was ordered and plans for sending the chosen Tuners to Switzerland for their training courses had been made. Everyone, down to the telephonist, was agog. By September affairs were far enough advanced for us to tell the Board that on Capital account of the new project we were likely to save £2065 on the £78,000 budget. The £6,000 overrun on the weaving shed being balanced by numerous small ups and downs and the non purchase of a twisting machine, and the decision not to extend the warehouse.

September 26 to February 27 went by without a Board meeting. In that period Sir Hugh Watson died and his passing was mourned by many, not least by me and Colin. His memorial service in St. Giles saw a full kirk, recognition of the regard in which he was held by young and old. The Board decided that the Chairmanship would hitherto be meeting by meeting on an alphabetical rota.

A 5% interim dividend was declared. It was noted that the weaving shed building was running six weeks late, but that it would be ready to receive the machinery. "The Colonel", and his wife Lydia, retired from the shop and Gardiner paid them for the goodwill which they had established.

114

1967, Centenary year. Not of the Limited Company, but of the firm's foundation; surely a time for celebration, especially as we appeared to be prospering. So we celebrated. Four parties were arranged. A buffet dance for employees and their attachments, which was held in the Victoria Hall and was a great success. We went to a lot of trouble in the provision of transport to ensure that nobody had to drive home. Only one person fell foul of the police, and he was hauled in for nothing worse than being drunk in charge of a bicycle. Then there was a dinner at Ednam House, in Kelso, for senior foremen, designers and managers, again with transport laid on. That one ended just before midnight when our singing was deemed in danger of upsetting the residents. A cocktail party at Wool House, the home of the International Wool Secretariat in London, for our customers, followed by dinner with our agents. And, last of all, dinner at Prestonfield House in Edinburgh for the Board, James Williamson, Deputy General Manager of the National Commercial Bank of Scotland, James Bogie and Jack Shaw, partners in Graham Smart and Annan, our Auditors, and Ken Smith and Gerald Box.

Archie Phaup, the mill's van driver and mechanic, veteran of the Burma Road, drove me and Colin up to Edinburgh. Potter and Earnshaw were flying up in the CPA corporate aeroplane. As we came over the top of Soutra Hill and saw the summer evening mist forming in the Forth Valley we wondered whether they would get into Turnhouse. They did, but they made the mistake of letting their pilot take the 'plane up to Leuchars to see his erstwhile colleagues at the Air Force base there. The fog prevented him from getting back to Turnhouse the next morning and Potter and Earnshaw had a dreadful taxi ride to Leuchars to get to their aeroplane.

Dinner, with minimal speeches was a great success. Lots of port rounded it off and, on leaving, Jack Shaw was given, by the manager, the flowers which decorated the hotel's hall to take home to his wife. To this day Shirley Shaw thinks that Jack stole them! Perish the thought; Sir John Shaw is now a Deputy Governor of the Bank of Scotland.

Meanwhile planning for the new machinery was almost complete. Shift working meant, in those days, male weavers and as the grade of assistant tuner was being phased out in the industry assistant tuners in the pattern department were offered, and accepted, the Sulzer weaving jobs. The change over meant redeployment of many workers throughout the mill and it is notable that the workforce of the time co-operated whole heartedly in the changes. Had the Union then the power it achieved in the Wilson years the transition to shuttleless weaving would not have been half so smooth. Basic minimum wage rates, holidays, hours of work, shift allowances and the like were centrally negotiated with the group of three Unions representing the workers in the Industry which were the Transport and General, the Municipal and General, and the Dyers and Bleachers, and it was at this time that the working week was reduced to forty hours with, as a quid pro quo, the abolition of the afternoon tea break. This, together with the introduction of shift working, brought about the installation of drink and snack vending machines in the mill.

Negotiations with these three Unions were generally friendly and rational, with most of the speaking being done by the Chairman of the Employers Association and the leading Trades Union Official. For many years that was Willie McBlane, of the G & MWU, a small man, extremely intelligent and a doughty bargainer. In the days when I was Chairman of the Employers he and I used to dispute possession of the one chocolate eclair which appeared at teatime during meetings. Each year at Christmas time the Unions and the Association staged a Christmas lunch, alternating payment and choice of venue. While I was Chairman Edward Aglen was secretary of the Association and he began the practice of having three speeches. Firstly his own erudite and supremely amusing ramble about the condition of the industry. Then the host's less competent exposition on some theme or other and then the guest's reply. It added point to the occasion and reduced the time devoted to the exchange of dirty stories which these events generate.

The oncoming spring season was expected to be difficult, and prices and overhead recoveries were pitched at a profit of £50,000. With £63,824 achieved in the year to January 1967 and greater production in prospect this was to prove conservative.

In March 1967 the eight Sulzer Weaving Machines arrived by rail at Galashiels station, were transferred to lorries and delivered to Selkirk. The drivers were able to reverse their vehicles along the half mile or so of railway roadbed from the level crossing to the eastern end of our new weaving shed and, the shed floor being just about at tailboard height, they were easily rolled in on bogies and set in position, glued to the floor. All the services had been built into channels in the floor so it was, literally, a matter of plugging them in and switching them on. Never in our experience did we install machinery which was easier to commission.

By April they were all running and we very quickly learned of the first error in our calculations. Our manning was wrong. Though operable, one tuner and one weaver per shift could not achieve the projected efficiency, so as soon as the men could be trained we moved up to two tuners and two weavers. This in fact represented marked overmanning, but in terms of product cost was more effective. We should have bought a first tranche of 12 machines!

The increased output brought problems and further need for reorganisation in its's wake. The yarn stock ran high, above budget, but was so effective that rather than reduce the stock we raised the budget. The yarn store itself was working out of the cartons in which yarn was delivered and the piling up which the increased stock necessitated caused us to order racking, roller conveyors and a fork lift truck at a cost of £7,000 to make the stock properly accessible.

Despite Gardiner's prosperity and the confident, expansionary atmosphere in the Company we still had serious problems with Executive Director's salaries. Or rather one problem. Our colleagues would not agree to increase them. In July this caused a Board Meeting to be held in Manchester whose primary purpose was to discuss the purchase of four more Sulzers at a cost of £25,000, this time of the more flexible dobby type. This was agreed, hire purchase being once again the mechanism of finance.

The arguments for such a purchase were not as solid as those for the first eight, and the Board paper does describe the proposal as something of an act of faith. With changes in manning in the Hattersley shed, and shutting down half of the looms and single loom weaving the rest, the objective was to move capacity up from 360 pieces per week to 400, which was the perceived level of demand. It is a comment on the confidence which we all felt that despite a poor trading background in the industry at large and the tentative nature of some of the arithmetic the Board agreed to go ahead. The secondary purpose was to discuss the salary question with the Chairman of CPA, then James Cullen, on hand if we could not agree.

By this time, after turning the Company round and restoring it's prosperity, Colin and I firmly believed that getting a reasonable reward for managing our own Company should be achievable without all the friction which every application generated. Had we not felt that we had reached some sort of sticking point we would not have trailed down to Manchester. When the talk of salaries began it was soon apparent that Geoffrey wasn't going to give an inch, so Mr. Cullen was asked to join us, and he proved to be just as intransigent. But we stuck to our guns, and after a long and difficult argument Cullen eventually stormed out of the room, telling Geoffrey to "give them what they want". Would that we had not had to raise the temperature so high to get a reasonable increase.

The half year to July 1967 yielded a profit of £31,970 and the Preference Dividend was paid. Indeed, it never again fell into arrears. The need for an independent Chairman was removed from the shareholding agreement. In December an interim dividend of 10% was declared and the year to January 1968 rolled on to end with a profit of £85,444.

In September, after extensive research into it's capabilities, we decided to instal ICT punched card data processing equipment to replace the Burroughs accounting machine and to take on board the growing task of production monitoring. Many other erstwhile manual tasks, such as cheque writing, were also planned for the machine which was installed in the summer of 1968 and went live, with a struggle, during the fortnight's holiday that year. Computers were not yet developed to the point where we could really consider them.

The CPA aeroplane had some effect on the venues of our Board meetings at this time. Geoffrey Potter was the moving spirit behind the purchase of that machine and took on responsibility for scheduling it's use. I have already mentioned the trip to the Centenary Dinner, and the next venture was even more exciting, when he decided to come to a Selkirk Board Meeting by air, landing on the patch of uneven common land high up by the golf course known as the Gala Rig, or the Selkirk racecourse, used once a year for the Common riding race meeting. Proposals had been made from time to time, notably by Harvey Wilson, to have it improved and designated as the Borders airstrip, but nothing had been done. Nevertheless, the CPA pilot, himself a Borderer, agreed to land on it, and they flew down on a day of low cloud which obscured the two television masts which stand within five miles of the racecourse. They managed to land safely, and got away again after the meeting.

Our March 1968 meeting was held in the Station Hotel in Carlisle, so that they could use Carlisle airport. The principal business was the decision to increase the Ordinary Capital by £60,000 a three for four bonus issue being the mechanism and making the shareholdings thereafter 70,000, 35,000 and 35,000. The reserves to be used were the Capital Redemption Reserve Fund, £34,000 the Investment Grants Reserve, £23,575 and £2,425 from the Profit and Loss account.

At that meeting, and at another held specially later in the same month an Executive memorandum on expansion was discussed and approved. No copy was appended to the minutes and no copy of that memo has survived, but later minutes reveal it's content. In order to push output well beyond the current target of 400 pieces per week four more Sulzers were authorised together with the revolutionary Hergeth warp sampling machine, two new twisting machines, new scouring plant, an additional Blowing machine and a new shop building so that the space it occupied at present could be converted to other use. The siting of the new shop is interesting.

Until 1967 the "mill lade" flowed between the Gardiner buildings and the Dunsdale Road, the water being provided by the cauld, or weir, which dammed up the waters of the Ettrick and Yarrow rivers just below the Selkirk bridge. Responsibility for maintenance of the cauld was shared among some of the Selkirk mills, Gardiner not being one of them, presumably because Tweed Mills had never used the lade as a source of power, or, indeed, for any purpose beyond poaching salmon and floating swans. In 1966 or 67, times being hard and major repairs to the cauld being needed, it was mysteriously blown up one night, allowing the rivers to flow freely again, depriving the lade of water and, many believe, so changing the flow above the weir as to undermine the foundations of the bridge, which fell into the river in a flood in the late seventies.

The noisome ditch into which the lade turned when deprived of it's waterflow was filled in by the Local Authority which then had the gall to ask Gardiner to pay for the land now created, within the wall which had bounded the lade and therefore, we believed, our property anyway. But they were adamant, and as it was the logical site for our new shop, we paid up, turning that part which was not used for the shop into pleasant lawns, a staff car park and greatly improved access to an extended cloth warehouse.

The year to January 1968 had shown a profit of £85,444 from a turnover of £1,039,951, the million exceeded for the first time. The half year to July 1968 yielded £73,320 so when, at the December meeting, yet more capital and expansion proposals were made they were discussed with enthusiasm and approved. The total proposed investment was £51,600 or £34,010 net of Regional Investment Grants, and included four more Sulzers which would bring the total of these machines up to 16. Two old Warp Mills were to be replaced by one new Hattersley RW4 and a new, extra, Broadbent 60" Hydro Extractor was included in the programme.

An extension to the weaving shed was required to house the new Sulzers. The Sim's building, north of us across the old railway line, part of which we already leased was, with it's field and a good three bedroomed house, offered to us for £5,000, and ultimately sold to us for less. And, finally, the existing Lancashire boiler, now oil fired, being within four years of a compulsory pressure reduction and struggling in the winter months, was to be supplemented by a Thomson Cochrane packaged, oil fired, boiler to handle the whole of the heating load. This was to make Tweed Mills one of the few two funnelled mills in Scotland!

The arguments for the proposals leant heavily on the continuing trend towards lighter weight fabrics which, effectively, means cloths with higher picks per inch with the concomitant reduction in output in terms of yards per shift.

Things were going well and the result for the year to January 1969 was to be a profit of £115,933. An interim dividend of 5.914% was declared. This was the first declared net under the new Imputation system of taxation, and the peculiar rate, which does not yield a whole gross number, reflects dividend limitation, by now in force.

An indemnity was provided to the Bank to cover them should one of our ICT produced cheques be tampered with. The reason for the perceived need for this was that as the tabulator could not write words we had programmed it to write the figures twice on the face of the cheque, and the Bank saw this as potentially risky.

Despite all this, true to form, the Board refused to discuss the salaries of the Executive Directors.

Chapter Fourteen

Shareholder Relations

While Geoffrey Potter and Reg Earnshaw represented the CPA interest the relationships within the Board ran in two distinct courses at the same time. On the personal level mutual respect and, if not positive liking for each other then an absence of active dislike kept everything on a friendly basis. But partly because our colleagues sometimes seemed to be batting to orders, and mainly because of the fundamentally opposed visions of the proper conduct of Gardiner held by the management and CPA, four separate strands of conflict ran through the whole period. Indeed they affected relations right up to the time that the Company was sold.

Beneath all the other conflicts was the fact that CPA did not understand and really wanted no part of the Woollen Industry and were not reluctant to put forward proposals designed either to get them out or to gain control. With the shareholdings balanced as they were Gardiner was not a subsidiary of CPA and the reality of effective control lying with the management was, to them, an intolerable situation.

Not being a subsidiary they did not consolidate our results. They called us an associate some of the time and a trade investment at other times. Either way they looked at the value of Gardiner as a function of the dividend stream, which was negligible. For our part Colin and I were intent on building the value of the business by ploughing back profits, improving and expanding the machinery and wanted to keep distributions to a minimum, at least until the day came when there was money in the bank.

This in it's turn led, we believed, to a CPA view that if our salaries could be held to unreasonably low levels we might crack, needing the dividends to plug the salary gap. There was a further prop to the tight CPA view of salaries in that they were notoriously poor payers themselves, and feared that if we did better than their equivalent level of management it might bounce back on them.

These four strands of conflict, wanting out, lack of consolidation, dividends and salaries were to be a constant irritation and distraction all through the twelve years from 1960 to 1972. They led to some incidents which, had they not been so serious would have been funny and to some exchanges, both verbal and in correspondence which were robust to put it at it's kindest.

Salary arguments were routine and annual but apart from the 1967 row in Manchester they had never got out of hand. But in 1964 the "want out" problem raised it's head for the first time since UTR had declined to sell their quarter to us in 1959.

In May of that year Geoffrey Potter suggested to us that the new Chairman of CPA (Cullen) was unhappy about the "untidy" fifty percent holding which they had in Gardiner and had enquired whether or not we might be inclined to sell to them. This was not, of course at all what they wanted and it did not take long for the discussion to be reversed, for they knew, for we told them, that we would not at this point be willing sellers, and the argument took it's natural and intended course. Would we buy them out? The answer to that was yes, if a price could be agreed.

In mid 1964 the net worth of the equity stood at about £100,000, or £1.25 (twenty five shillings – old money) for each of the 80,000 Ordinary shares in issue. The price at which the holding was eventually offered to us was thirty shillings a share which meant that we would each have had to find £30,000. We knew that the holding stood in the CPA books at something like £37,000 that the beginnings of Gardiner's success were our doing and not theirs, and that, the Textile trade being what it is, those beginnings were not yet a secure foundation, so we said no. Would that we had said yes. But who knows; the discipline of having to justify our expensive expansive proposals to our Manchester colleagues may well have kept us from error!

The next couple of years were devoted to pushing on with the expansion of the business until in March 1968, at a Board meeting which approved a further Capital equipment programme Reg Earnshaw raised, in a serious way, the dividend question, and when the minute, when written, failed to refer to it he sought to have the minute amended and expanded in quite dramatic terms. His actual suggested wording, which was not adopted, encapsulates many of the CPA attitudes to the way in which Gardiner's affairs were being conducted. It reads:–

" Mr. Earnshaw explained the difficult position of the shareholder whom he and Mr. Potter represented, in that the constant reinvestment of almost all the profits earned in the Company resulted in what appeared to be a grossly inadequate return on the sum which could be realised by that Shareholder on the sale of his shares on an open market. If the Calico Printers Association Limited held a number of investments of this kind showing only this return it could only have the effect of depressing it's own shares to a dangerous level. Whilst he had no doubt that the plan as outlined was in the long term interest of Edward Gardiner & Sons Ltd. the Board of the Calico Printers Association Limited might object to the constant ploughing back of virtually the whole of the profits earned. He thought, therefore, that in consenting to the proposals it should be understood that it was the intention of all Directors that, as soon as the situation permitted, we should begin to pay out dividends at a level which would be equal to, say, 5% on the then value of the shares, that value being calculated without reference to any apparent value restrictions arising from the peculiarities of the shareholding agreement."

This suggestion fits neatly into a discussion then current, another proposal that we buy the CPA shareholding, which argument had by then, once again, got down to price. It seems from the correspondence, for we were not yet in the habit of filing notes of telephone conversations, that against an asset value of a shade over £3 per share we had offered £2.50, which would have cost us each £50,000 and yielded CPA £100,000. The negotiations foundered, but we did agree to devote time at the next Board meeting to formulating a dividend policy. That meeting took place on 22 July 1968 and reached useful conclusions.

On dividends the CPA minimum objective of a dividend of 5% of the value of the Company was accommodated by assuming a price earnings ratio of ten and agreeing to distribute half the net profit. This would, were all the assumptions valid, yield 5% of the value of the shares. It had to be recognised that this would not be practical until dividend restraints, currently Government policy, were eased.

The Board then turned to the longer term future of the Company, or rather the amendment of the fifty fifty shareholding position. No suggestion that CPA buy out the management was made this time, and the alternatives discussed were, firstly, that at some time Brown and Jackson buy out CPA, an idea unlikely to succeed in view of entrenched positions on price. Secondly, the possibility of finding a purchaser for the whole of the equity and, thirdly, the possibility of going public to the extent of 40% or less of the equity should the profit record permit such a course.

Colin and I looked askance at options one and two, though recognising that a bid for Gardiner was a possibility at any time, and the Board agreed that the third option was the most appealing and decided to keep the whole matter under review.

In the nineteen nineties it is difficult to remember that in the nineteen sixties firms were going public on profits of less than £100,000. Otherwise consideration of that option would have been a mite unrealistic, for Gardiner had not yet cracked the hundred thousand profit barrier.

In July or August 1968 English Sewing Cotton Ltd., having been forbidden by the Department of Industry to merge with Courtaulds, took over CPA and the whole became known as English Calico Ltd. This involved a change of Chairman, Neville Butterworth, of the dominant English Sewing, getting the job. As the Chairman of CPA had done when he discovered the existence of the Gardiner shareholding, so did Neville Butterworth waste no time in putting his oar in. His particular line of attack was the thought that whilst at present was an unconsolidated trade investment or associate, with 50% plus one share it could be a consolidated subsidiary and his representatives were despatched to the December 1968 Board meeting with a private agenda.

The main business of the meeting was consideration of a proposal to buy four more Sulzers and sundry other machinery at a net cost after grants of £34,010. That, and a dividend limited payment of 5.914% were swiftly agreed, and there the minute ends. But after the experience of Earnshaw's attempt to get expanded inflammatory items inserted, and perceiving the acceleration of the dividend, buy-out, salary and one share arguments we took the trouble to make a private note of the other events of the meeting.

First we endured an attack on our return on capital, which was abandoned when the arithmetic on which it was based was shown to relate a half year's profit to full net worth. Then, in a discussion of the salaries of the Executive Directors our request for a periodical review was turned down on the grounds that Mr. Butterworth was interested, and wished to meet us and it could wait till then.

At that point the Potter/Earnshaw double act got into it's stride, with Reg suggesting that a major cause of friction was unconsolidatability, and that if only we would sell one share, or issue one share of a new class with no voting rights, the problem would be overcome. No discussion of this idea took place, for Potter said that he wasn't in a position to do so as he had never heard the idea before, a statement belied by some of his subsequent comments. But he did opine that we should think about it and "if we did not want to fall in with the suggestion we need never mention it again". Our note continues "This may be important for I feel sure that CPA will return to the subject and the idea from our point of view is not a very good one". The note goes on to list the disadvantages, from our (me and Colin's) point of view which were, briefly:–

1. Our image. Should we try to sell Gardiner it might be less easy as an acknowledged subsidiary of CPA

2. We might become involved in CPA's woollen division, which meant the worsted weaver, Heywood.

3. The way might be open for CPA to claim that as a subsidiary we should allow inspection by them of anything they chose to inspect.

4. Involvement with their salary structure.

5.   Involvement in their consolidation procedures, deadlines and accounting practices, as well as detailed forecasting, for which they had been pushing already.

6.   Whilst voting rights at an AGM might be clear, a Board meeting of a subsidiary might be different from that of an independent Company.

7.   Agreement would exaggerate the tensions which already existed.

And so, at Neville Butterworth's suggestion, on 3 April 1969, Colin and I went to Manchester to meet him. It was to be a famous meeting. The programme was for a chat and discussion on business philosophy in the morning, lunch in the private dining room at Arkwright House, a discussion on salaries and home for dinner.

The first part went to plan. We met, had the ramifications of English Calico described to us and were then treated to a lengthy exposition of Neville Butterworth's personal business philosophy as applied at English Calico. We managed to get in the odd word, and did our best to promote our own philosophy, but it was swimming against the tide. Then lunch. Lunch at the old CPA building across the other side of Oxford Road had always been a sumptuous affair, and English Calico had stressed their lesser pretensions, but the down grading was barely noticeable. There were present Neville Butterworth, at the head of the table, Potter, Earnshaw, Colin and me and Tom Wetherby, the relatively new Finance Director of English Calico Limited (ECL).

All went smoothly till near the end of the main course, when Butterworth launched into a tirade in which he stigmatised us as unreasonable for not falling in with the "transfer one share" suggestion. We could have a blank transfer, signed, so that we could get it back at any time, no change in the operation of Gardiner was contemplated and things would go on just as before, but ECL would be able to consolidate Gardiner.

We responded by listing our objections, wondering why, having regard to the relative sizes of ECL and Gardiner, consolidating our modest profits was a matter of any moment. By the time we got to the sweet tempers were rising. The explosion came when I said "and anyway, you can consolidate us now, with only fifty percent; it is

127

common practice in Scotland, and I'm sure it is done in England too". At which Neville Butterworth, clearly absolutely livid, flung his napkin into his fruit salad, strode out of the room and never came back. We drank our coffee, took our leave and left. We never saw Neville Butterworth again. The discussion of salaries never occurred.

During the day there had been some talk of finding a purchaser for the ECL half, and on our return to Selkirk we wrote seeking sight of their consolidation schedules and paper work, the detail required in their profit forecasts and anything else which might fall upon us were we to be consolidated. We also asked what price they would want were we to encounter a potential purchaser. And we mentioned salaries, and ECL practice in that connection.

The reply has gone astray, but subsequent correspondence shows that no progress was made on any front and, in particular that an independent professional appraisal of our salaries was to be ignored.

The paper war continued, with February and March of 1970 producing exchanges of long letters demanding forecasts and using the disappointing result for the year (it was to be £76,436 against £115,933) as a base from which to attack our policies and our conduct of affairs. "This vast investment programme, necessitating the retention of all profits earned over a considerable period of time, those profits being pulled down by high overdraft interest charges, has operated wholly in your favour as individual shareholders, and is causing us increasing embarrassment as time goes on". That letter went on to savage our management of stocks, debtors and creditors and suggest that we should follow the then developing big company practice of deferring payment to suppliers beyond the agreed date to reduce working capital.

Our response was equally forthright. After listing the profits since 1964, culminating in 1969 at £115,933 representing a return on capital of 36.4% we went on to say:–

" Furthermore, whilst the unconsolidated profitable operations of Gardiner may be an embarrassment to your group your continued suggestion that Gardiners is operated solely for the benefit of myself and Mr. Brown is quite uncalled for and, unless the conversion of a company which as an investment stands in your books at, we believe, a figure substantially below par, and which company was near extinction, into a profitable concern is regarded by you as undesirable, has no foundation in fact.

You appear in your letter to question the wisdom of the Board's policy over the years in its having invested in new machinery. I would suggest that you obtain a copy of the Atkins Report on "The Strategic Future of the Wool Textile Industry". This was published by our industry's NEDO and you will find on reading it that not only does the investment that we have made anticipate the report's recommendations for the industry, but so also do the commercial policies which we have adopted anticipate the expected future pattern of our industry".

This produced an emollient reply, but nothing had changed.

The disagreements were, by April 1970, taking the form of a complete refusal to discuss our salaries. And, from our side, a reluctance to contemplate the payment of a second interim dividend: things were looking tough, the overdraft, as ECL had so rightly pointed out, was rising and a dividend would represent more borrowing and even higher interest charges.

We refused to go to Manchester for a meeting to discuss both subjects, and Potter and Earnshaw refused to come to Selkirk. So we compromised, they flew up to Carlisle in their aeroplane, and we drove down, and the meeting was held in the airfield manager's office, a garden hut in the middle of the field. We were well prepared, and knew that under the Company's Act the declaration of a dividend is a positive act, that is to say that whatever the Directors propose it requires a shareholders majority to pay it. An interim can be paid by the Directors without reference to the shareholders, but with the Board split 2 to 2, with the Chairman having no casting vote, no dividend could be paid or recommended unless Colin and I agreed.

It was a shortish meeting. The ECL men offered £500 on our basic salaries which, in the light of the professional report in our hands, we refused, and we declined to sanction a dividend.

This must have been the nadir of our relations, for the next week, in telephone discussions, we agreed to commission Hay-MSL to report on our salaries, and to abide by their recommendations.

From then on the tension eased. The report proposed salaries which even we thought too high, so agreement was easily reached. The general agreement cannot have been based on our performance because the next three years were tough and showed declining profitability, and the improvement was to await the retiral of both Potter and Earnshaw before it showed. Outlandish ideas still came out of the ECL machine. In November 1970 we were asked whether we would approve of Butterworth, by then Sir Neville, trying to persuade one Partridge, Chairman of Imperial Tobacco, to shut their wholly owned subsidiary Gibson & Lumgair, a bitter rival of ours who were losing money in bucketfuls. Then, later, and just after his retirement, Reg Earnshaw was put up to suggest that Gardiner bought Laidlaw of Keith, a prosperous rival in the North, a suggestion which for many reasons held no appeal for us. In neither case did ECL press the point.

Geoffrey Potter retired at the end of March 1972 after having made sure that his successor Bob Audsley was indoctrinated in the ethos of and the peculiarities of the Gardiner situation. He must have done it well, for peace reigned thereafter. Till 1975.

Reg Earnshaw retired in September 1972 and was replaced by Norman Hornsby, who knew something of the Scottish Woollen Industry, having once managed D. Ballantyne Ltd. at March Street in Peebles.

Both retirements were marked by dinners in the Borders to which the new Directors were not invited. Old Boys only.

Chapter Fifteen

The Honeymoon is Over 1969/1972

In textiles nothing lasts forever, or even for very long for that matter, and the year to 31 January 1969 was to be the best for some time. In the three years that followed profits were £76,436, £28,562 and then the nadir of £14,436. All these profits are reported after adding back the bonuses which the Executive Directors took instead of their dividends.

Fashion was changing. The cloths upon which the reorganisation had been founded were less wanted, and lighter cloths of single yarn construction and taking longer to weave were being demanded. The structure of the garment making industry was also changing, with hitherto great names, important to Gardiner as bulk consumers, declining in the face of the advent of the boutique concept, both in terms of shops and their suppliers. Whilst the fabric change was progressive and apparent the structural change in garment making had yet to become obvious.

Looking back from 1991 two basic facts about the industry seem plain. Firstly the operations of a textile company must be completely overhauled every ten years or so, the review being comprehensive and probably involving major product changes and substantial re-equipment. Only companies willing and able to put their operations under scrutiny in depth can hope to survive.

The other fundamental concerns the Scottish Tweed Industry as a whole. If there is one single factor which underlies it's decline to the present remnant of it's former self it is the failure to go fine enough fast enough. The demand for lighter fabrics demanding finer yarns and higher picks per inch was matched in the spinning industry which was being asked for finer yarns not only by weavers but also by it's knitwear clients. The reasons for the failure are not far to seek, for every step on the road to fineness reduces the output of existing plant, even if it is capable of the change, and so demands the installation of more plant to bring output up again. And machinery, as it became more and more sophisticated increased in price much faster then the general inflation rate.

It was against this background that Gardiners had to face up to the ending of the run of success which had flowed from the High Straight decision, now nearly ten years old.

The downturn which was upon us was not immediately apparent and when the January 1969 accounts were reviewed in March the Company, with 286 employees, later to peak at 330, was the largest employer in Selkirk. The recently founded Exacta printed circuit Company had not yet overtaken us. After reviewing the accounts the Board agreed to buy a Savio winder to wind the weft package for the Sulzers, large enough to service the sixteen we had already which were about to move up from ten shifts per week to fifteen, running round the clock Monday to Friday, and four more not yet officially contemplated. The Stubbs winder already on site was not performing well.

The progress of the last few years was London womens wear based, and therefore Alan Shepherd based. Ken and Gerald had made noises about their desire for an agreement which would protect them from dismissal without compensation and we agreed to make one, at the same time protecting ourselves against their taking on conflicting agencies. There were no difficulties in this , and by July the agreement was in place.

By then the downturn was making itself felt even though the half year was to yield a profit of £74,608 against the previous first half of £73,230. With a profit and loss account balance of £175,000 there were no impediments to paying the Preference dividend, but nothing was suggested for the Ordinary shareholders.

To the north of the Gardiner ground, across the bit of railway line which we had bought and on part of which the Sulzer shed stood was the semi derelict Palfrey's, or Sim's mill, part of which we already leased for storage purposes. In another part Turnbulls of Hawick carried on a very small winding operation which they now shut. As the building was now of no use to them they offered it to us, along with 3.3 acres of land and a good three bedroomed detached house, all for the princely sum of £1,275. We accepted, and thereby made our space for expansion, for hitherto we had, apart from our railway venture, been completely hemmed in by our neighbours.

Gardiner's fortunes were slipping away, for the second half, to 31 January 1970, made a mere £1,828. Even so, though morning mail openings were taking longer because of the agonising about what was wrong and what had to be done to correct it we had not lost confidence, and in March ordered another Boyd 100 spindle twisting frame suitable for our continuing assault on the mens wear trade.

The second year of the slide, to January 1971, which was to yield a first half profit and a second half loss of £538, was a year in which the first steps to get things back on track were taken. Back in 1969, looking for the possible next generation replacement for the Hattersley looms used for pattern making, we had bought one Snoeck rapier loom, believing, incorrectly, that it's setting up time would be nearly as short as a Hattersley's and nothing like as long as the three to four hours required by a Sulzer. Though it did not have the advantages for which it had been bought it did have an eight weft colour capability compared with Sulzer's four. Tartans were in fashion and we already had part of the Gor Ray contract, which was substantial, but uneconomic if woven on Hattersley looms, which it mostly had to be for the bulk of tartans have more than four weft colours. Not only did we have that contract, but the great Scottish chains of tourist shops were developing rapidly, led by Moffat Weavers, who wove only for the entertainment of tourists, closely followed by Pitlochry Knitwear, who knitted not at all, both very soon outpaced by the Edinburgh Woollen Mill. Lesser players were Murray Brothers and Highland Home Industries, later to become Clan Royal.

All these chains used large quantities of cloth. Moffat began by selling lengths of cloth off the roll, both tartan and traditional Scottish tweedy designs, and we had been the supplier of the latter for a number of years. But at the beginning of the seventies they all moved away from selling cloth to selling garments, and as our tweedy business with Moffat declined the opportunity to get their tartan contract presented itself. To do this, and to make the Gor Ray business worth while, we bought three more Snoecks to make a unit of four. These were ordered in July 1970 for delivery in February 1971. At the same time two 120 spindle Gemmil and Dunsmore large package ring twisting frames were ordered to replace the ten and less year old two for one machines. This change was partly inspired by the Equal Pay Act, which looked as though, through time, it would push up women's wages, and the assembly winding operation which preceded two for one twisting was female labour intensive. All this involved re-siting the twisting department and relocating some other machinery.

Quite clearly there were a number of problems to be resolved and not too much time for discovery of the solutions so calendar 1970, the second year of declining profits, was largely devoted to thinking about the future, thinking which involved our agents in meetings, with us to discover not only their views of Gardiner's present state, but their thoughts on whatever changes might be needed.

Experience had taught us that a rambling open ended discussion would be unlikely to produce solid conclusions or recommendations and that unless the underlying facts were known to all before the talking started all that would result would be a very tiring day. So in preparation for a July conference of Colin and me, Ken and Gerald and Jack Phillis and Clive Guest we prepared a document setting out the position as it then was, and asking 35 questions which should be addressed on the road to a conclusion.

The meeting duly took place. It WAS a long hard day. It did not produce the touchstone but it did crystallise everyone's thoughts and did lead to two further pamphlets dealing with what conclusions we had reached and tentatively floating the idea of making quantities of our best designs on spec without orders, an idea which came to be known as "make and sell". These papers led to no further conferences but were well hashed over by us all during our normal, frequent meetings.

In the end "make and sell" in its proper sense was never instituted, so we will never know whether it would have succeeded. But in it's restricted form which became known as instant sampling it did become part of our policies. At the beginning of each season for many years thereafter half pieces, or in some cases pieces, were made of a considerable selection from the range. From these sample lengths could be cut and sent immediately to customers with the added advantage that if ordered twice an item did not have to be made twice which did markedly reduce sampling costs.

This period was the peak of our trade with Marks and Spencer who, at that time, did an enormous business in pure new wool skirts the cloth for which was provided by Gardiner, Brown Brothers in Galashiels, Gibson and Lumgair in Selkirk and, latterly, Laidlaw of Keith and Hunter of Galashiels. The head of the appropriate buying department at Marks and Spencer was Terry Coleman, ably assisted by Maureen Crabbe who had a soft spot for the Scottish industry having trained at the Scottish Woollen Technical College. It was a massive operation involving thousands of pieces each year and was tightly controlled. After sampling Marks and Spencer placed their bulk orders at the back end of the year or in January, for readiness from April through to July or sometimes August. There was an elaborate system of reporting, inspection visits by the department not only to check on quality but also to see that the production reports were not fairy tales. The rate of offtake of cloth depended on the efficiency of the garment makers to whom it was allocated and to the buoyancy of sales in the stores. In some years the cloth was called as fast as it was made, and in others it became a battle to have it taken in. In one particularly poor year some of it had to be carried over and Marks and Spencer did agree to make loans to cover the cost of the fabric which could not be delivered.

On one occasion, a year when cloth was being taken at a reasonable pace though not "hot off the loom" we were a little behind schedule and our production reports were slightly economical of the truth. During August, the tail end of the season, we did not dare to reveal the scale of the slippage lest Marks and Spencer be inspired to cancel part of the order and cloth was being reported ready for which we had not yet got the yarn. Imagine our horror when Pogson rang up to say that he had had a fire, and the yarn concerned had been destroyed. Somehow or other we escaped without being discovered. On another occasion one of the inspection visits was rather spoiled, for on the day of the visit the weaving shed was stopped by a strike resulting from one of our periodical fights with our Tuning staff.

As with everything else in textiles the woollen skirt boom did not last forever. It gave way to synthetic skirts, of which we had one years orders, but which became varied as to fibre and drifted away from us so that in the late seventies and early eighties our whole Marks and Spencer Trade was in a pure new wool coating for children.

In the latter years of the woollen skirt came a development to which Gardiners took exception, but which we could not control It was this. In order to keep a number of cloth suppliers involved Marks and Spencer would make their selection of design, and then farm it out around their cloth suppliers, so that we might get an order on Brown Brothers designs, and they might get one on ours. As it was chiefly our designs which were chosen we thought we were providing a free service to the others, but it did work in the other direction sometimes. And we did, always until the wool skirt fell out of favour, get as much as was good for us. Not as much as we wanted, but as much as was good for us.

Copying of designs was commonplace in the industry, copying of cloths less so, but in connection with Marks and Spencer and the attraction that their business had for cloth manufacturers I came across the copying incident which did more harm to my blood pressure than anything else I experienced in my years in the trade. Sitting peacefully in my office one day I got a telephone call from a man, who shall be nameless. He was a young man, and I and others had tried to smooth his path in the trade associations, getting him onto committees and involved in affairs. He was inclined to verbosity, but what he said can be condensed into: "I am about to copy one of your cloths for Marks and Spencer. How much do you charge them for it?" The conversation did not last long.

In September 1970 the Board slipped a cog and paid the Executive Directors a bonus of £3,150 each, "in recognition of a first half profit in exceptionally difficult circumstances"!

At the end of October, when the July to September figures were discussed, it was noted that the whole of the sales shortfall was in

London and the idea of selling through our own man without agents raised it's head again. We were instructed to formulate a plan for doing that, not to implement it but to report back. This eventually, the following year, turned into a plan to buy the Shepherd agency and employ Ken and Gerald directly. Initially they were enthusiastic and had the first plan not been thrown out by our tax advisers it might have happened. By July 1971 their enthusiasm was waning but another package was prepared, but November's minute records that the purchase proposal had been terminated.

November 1970 saw the stopping of 25 Hattersley looms, the best of them being transferred to the pattern shop and the rest cannibalised for spares or scrapped. 27 redundancies, our first batch, resulted, costing £620, for in those days the state picked up half the cost. This was partly the result of the Sulzers having moved to three shift working, partly because by February we would have four Somets in production but mostly because of a combination of recession in the trade and the growing criticality of the spring season, for which no solution was in sight. By November 1971 the state of trade and Gardiner's policies led to an increase in prices higher than would normally have been the case in order to allow bigger discounts to the largest customers, such as Marks and Spencer. The days of slapping the price list on the table on a take it or leave it basis were long gone. And, in particular, special pricing of the mens wear line, which sold mostly in the spring season, was authorised as one of the ploys to deal with the problem.

Since the reorganisation of the Company in 1961 ("High Straight") the whole operation had been home trade oriented, with minor sales being made in North America, the Far East and Holland. No other export market made any contribution worthy of note and indeed Export sales had in the period 1961 to January 1970 never exceeded five percent of the total. Hence our obsession with getting the Home Trade back on track. But early in 1970 Colin became aware of the substantial presence in the market place of garments made from cloths similar to ours mainly from Irish mills and carrying labels "Styled by Wilhelm Rohe" and, through his contacts with Peter Finlay of the Merchant Harding and Messenger, who knew Rohe, learned that besides being a stylist Rohe also represented a number of mills as selling agent in West Germany. Since our present agent in that country, Arthur Ernst, had not been terribly successful and was, unfortunately, terminally ill, Colin persuaded Peter to arrange an introduction.

The meeting took place in July 1970 and Colin and Bill Rohe took to each other right away, yet another unlikely partnership for if they shared one characteristic alone it was a stubborn streak, which they managed to keep under control sufficiently well to let them work together with tremendous effect. All business, in whatever field, is much more about people than about finance, or machinery, or even, dare I say it, design, and in Bill Rohe we had found one of those people who make things happen. And it was fortunate that we did, for whilst the Home Trade languished and the Spring Season problem became annually more serious, exports through Bill Rohe were to plug the gap for many years to come. The evidence is in the statistics.

Annual exports as a percentage of turnover, up to

1970	5%
1971	15%
1972	24%
1973	38%
1974	33%
1975	34%
1976	40%
1977	50%
1978	38%
1979	29%
1980	22%

All years to 31 January

In the growth period of the export percentage the increase is, if not entirely, then largely, attributable to Bill Rohe, though in the later seventies Canada and France developed into useful markets. The decline from 1978 onwards has special causes which will be explored later.

Rohe was no ordinary agent, for he conducted a merchanting business as well which not only gave Gardiner a share of a market hitherto untouched but also allowed him to conduct a vast business from a relatively small base of pattern ranges and sample lengths. Twice a year he would spend some days in Selkirk indicating his design requirements for the coming season and then, when the ranges were available he would order a large quantity of sample lengths, pieces and half pieces, which would form the basis of his merchant activity, and also provide sample lengths to the agency, or bulk customers. In this way he was able to cover a large field most economically.

The year to 31 January 1972, with it's profit a mere £14,435 was not one to remember. We had no reason yet to know the dramatic impact that West Germany was to make on our fortunes, but when the accounts for the year were considered at a meeting on 27 March 1972 the Home Trade was already looking better and we were disposed to think of moving forward again and so more machinery purchases were approved.

Our mens wear operation was bearing fruit, partly in thornproof cloths which had a high twist content, higher than our present capacity could handle, so we took the chance of a seven year lease, with option to buy, of a Berwickshire County Council factory on the trading estate at Eyemouth. To it we moved our Eyemouth darning staff, and in it we installed two second hand twisting frames bought from a Monsanto subsidiary, John Shaw (Dry Spun) Ltd., in Stainland near Halifax.

We reviewed the poor performance of the section of four Snoeck rapier looms, which was partly because the machine wasn't as good as we had thought, and partly because the setup of a separate section of four non Sulzer machines had not been a success, bringing all sorts of bickering in it's train. But help was at hand, for Sulzer had just brought out a machine identical in all important respects to the sixteen we already had but permitting the use of six weft colours. With a little manipulation six colours would make the vast majority of Tartans even if (when) the delightfully simple three colour Black Watch ceased to be the market leader. So we agreed to research the machine and, should the result be favourable, to buy four to replace the Snoecks. This would need yet another extension to the weaving shed and, after the investigation, the machines were bought and delivered in 1972.

It was to be Geoffrey Potter's last meeting. As a consequence of reorganisation following the English Sewing/CPA merger he was to resign on 31 March 1972.

We were about to bounce off the bottom for the second time.

Chapter Sixteen

Quiet Passage 1972/1974

Quiet is relative, and means more that Gardiners were not involved in substantial new ventures than that the conduct of the business was easy and peaceful.

The two calendar years 1972 and 1973 saw a dramatic shift in the emphasis from home trade to export, and within that shift a major realignment of home trade operations. The spring season problem became more acute but effective solutions, in the event temporary, were found. They were years in which, though a Conservative Government under Edward Heath was in power the Union representing the workforce, the T & GWU flexed it's muscles on the strength of the previous six years of Labour Government. It was a period in which the conduct of business was hampered by both price control and by a wage freeze followed by strict wage limitation, neither of which policies made running an expanding business in a constantly changing industry any easier. And dividend limitation did not improve shareholder relations, the agreed half of the net profit wildly exceeding the amounts which we were permitted to pay.

Exports flourished, led by Bill Rohe in West Germany and, as indicated in an earlier chapter, they did so at minimal cost in sampling and at margins improved by a weakening pound. Capacity in the first half of the year became a constraint and an elaborate system of space reservations for customers and markets was set up which, like all such, was more honoured in the breach than in the observance. The shift of emphasis to export accelerated the move to finer single yarn cloths as opposed to twist fabrics and forced the expansion of the weaving plant to accommodate this trend. At the beginning of the period there were sixteen Sulzers and four Snoecks in service, all running three shifts. In 1972 Sulzer brought out their six colour machine and, in view of the unsatisfactory performance of the Snoeck section, four of them were ordered for delivery in December 1972. Tartan demand was such that the Snoecks were kept going until early 1974, when they were sold for £9,500 net of the leasing company's penalty.

The year to January 1973 saw a profit revival to £100,763, the second best ever, and demand for the product was still rising so six more Sulzers, of the four colour dobby type, were ordered for delivery in January 1974. Inflation and the weakening pound were taking their toll and whereas the first four dobby machines had cost £6,500 each in 1968 the 1974 machines were installed at £13,750 apiece. And they required yet another extension to the weaving shed.

Other capital additions in the period were the installation of forced ventilation in the Pattern Shop, Twisting flat, the Sample Warehouse and the Milling House, all in late 1972. The following year saw a new van, yarn conveyors and a flock collection plant for the dry finishing department, and a warning that the Lancashire boiler would very soon have to be replaced. With the expansion of weaving capacity yarn storage was becoming a problem and a new yarnstore, sited on the railway line immediately to the west of the weaving shed, between the old yarn store and the Palfrey building, to be fitted with powered racking to hold 100,000 kg of yarn, was approved. This was not built in that format, the building being made larger and the powered racking abandoned. The purchase of a decatising machine and associated weft straightener was discussed, but the project was shelved, to be revived a few years later. Such a machine would allow Gardiner to substitute a finish of their own in place of the "London Shrunk" finish provided by John Gladstone and more and more demanded by garment makers, particularly in Germany where mechanised production demanded minimal movement of fabric in making up and pressing. Though this was a relatively quiet period capital expenditure scheduled at 17 July 1973 amounted to £141,000, and much of it was destined for leasing or Hire Purchase.

The ICT punched card equipment had been thumping away since August 1968 doing all the accounting work and yarn stock and work in progress control, but by it's very nature it lacked the calculating power which was now available in the smaller computers which were becoming available. A move to one of these machines to take advantage of their power to calculate and print invoices and to handle the whole of the payroll function as well as doing all that the existing equipment could do seemed appropriate, and the Burroughs 2000 machine seemed to be the right machine, so an order was placed for one of these for delivery at the 1972 summer holiday.

This machine would take us away from card input to direct keyboard input and would reduce the amount of paper output by returning, in many functions, to a ledger card presentation of data. Programming would take months and months so, in the summer of 1971 I went off to Southend to learn how to specify a programme in a form from which the programmers could work. Returning, I set about my task, specifying the yarn stock control programme first. A few weeks later the programme was in Burroughs hands in Edinburgh and I took my test data up there to run it on their machine. The programme worked beautifully, but it took only one printing run to make it clear that the machine could not handle even the yarn stock, let alone all the other functions required of it. I had misunderstood the salesman's patter, and believed his figure of 100 characters per second printing speed, not realising that that speed could only be achieved if no calculation function was involved. There was nothing for it but to cancel the order. Burroughs were persuaded to pick up the bill for the programming done so far. The salesman was exiled to Tanganyika or some other African state and we prevailed upon ICT to leave the punched card kit in place for another year.

Tootal had used ICL 1901 machines for many years so I went and inspected their set up, though I truly believed that the machine was a bit on the big side for us. I came away sure that it would do all we asked it to and resisted all persuasion to instal a terminal at Selkirk connected to the Tootal machine. We ordered a 1901 k for start up in August 1973 over the summer holiday, commissioned an ICL subsidiary to write the programmes and sat back and waited, providing the data as required but, experience showed, not taking nearly enough interest in the programme testing effort.

The August start up was fated from the moment the lorry bringing part of the equipment up to Selkirk burst it's radiator on the M6 and in the cloud of steam ran off the road. Everybody connected with the installation took holidays early so as to devote the fortnight to getting the system up and running before the mill reopened but that was not to be. After a week of desperate effort the programmers conceded that bugs were appearing faster than they were being cured and the start up was abandoned. Fortunately the punched card machinery was still in place so we were able to start it up again and run it till October when a second attempt was successful.

In the earlier years of the reorganisation no grants for capital expenditure were available, the only state contribution being through tax allowances which ran up eventually to 100% allowance in the first year, which almost made Corporation Tax a voluntary levy. But in the late sixties various categories of development area were devised which ranked for differing percentages of cash grant against capital spending. The first benefits accrued to Gardiner in the year to January 1966 when £1,200 was received against a capital spend of £15,728. Thereafter, down to 1980, most equipment and new building, with the exception of motor cars, earned grant. In the early days the grant hung on creation of jobs, but later it became automatic, and the heyday came after 1972 when, besides the standard Regional Development Grant (RDG), sums provided by Section 8 of the Industry Act 1972 became available. Like the RDG these were at the rate of 20%, but depended on the scrapping of equivalent plant capacity. So, in the period under review the ultimate cost of capital plans was reduced by 20% or more. Over the whole grant aided period from 1966 until the final contribution was received in 1982 Gardiner got £643,606 on purchases of £2,838,561, all of which grants ultimately flowed through to the profit and loss account.

The home trade was moving north. The majority of the London and Leeds garment makers had failed to mechanise and keep up with the times and Gardiner's trade with them was declining, quite apart from the fact that many of them were departing the scene. Great names like Mornessa, W & D Peake, Crayson, Harella, Freddie Stark and London Maid ceased business at about this time. Marks and Spencer remained the most important customer, still using woollen skirting cloths in bulk. But the major developments were occurring in Scotland.

In the post war years many mills, including Gardiner, had opened a mill shop to sell cloth, scarves, ties and the like to locals and tourists.

In most cases it was a case of having an outlet for getting rid of mistakes, very much a peripheral activity, but a few men of foresight realised the potential of the tourist trade and began to structure the mill shop to take advantage of it.

The first major player was Moffat Weavers, a small firm running a few handlooms in Moffat in Dumfriesshire. Blessed with space for a large car park they began to attract bus tours to their establishment, to watch the hand weaving process and to buy "the product" in their shop. More and more as the trade developed was the product bought in from Scottish weaving mills and Gardiner became a principal supplier of traditional tweed and, more importantly, secured at about this time the contract to make tartan for them; tartan which they both sold as lengths of cloth and, more and more, as made up skirts and kilts. Not only was this a very large business which, at it's peak with Gardiner amounted to 1600 pieces in a year, but it was a spring season stopper. The nature of the trade was such that demand, tartan by tartan, could be predicted with some certainty and by hard negotiation with David Murray, the Managing Director of Moffat who had succeeded Frank Scott, the founder and originator of the idea, Gardiner were able to turn that to advantage. As the business developed the contract for the year would be placed in August, with a colour specification of more than half of it immediately, for readiness from November through to January. The rest of the contract would be coloured later in the year, for delivery through to May, which meant that a large part of it went to plugging the spring gap. It was this contract that triggered the eventual purchase of a decatising machine, and the relationship with Moffat was a great success for many years. But nothing lasts forever and one year the yarn supplier let us down, delivering weak yarn late, and we fell down on our delivery promises. The following year Gardiner got only half the contract (the other half going to the weaving division of our yarn supplier), and soon after that the cloth changed to a worsted fabric on which we were uncompetitive. It was great while it lasted.

The next major player on the scene was Pitlochry Knitwear, a firm with whom, not for lack of trying, we never managed to do really substantial business. Perhaps that was not a bad thing, for the tourist operators were incredibly jealous of each other and it may well be impossible to deal with all three of the major companies in a substantial way at the same time. With two out of three Gardiner perhaps did as well as it could reasonably expect.

Before dealing with the Edinburgh Woollen Mill, the biggest tourist operator of them all, Angus Macintyre of Inverary, our first serious tourist customer, must have a mention, for he was our first customer in this field. It took a year or two, but we managed to move from selling him bits and pieces of stock to between twenty and thirty pieces a year at threes to a colour. The annual visit to Inverary was a very pleasant day out, driving there by way of the Gare Loch and Loch Long and Loch Fyne through the glorious highland scenery, doing a little business and coming home again. Angus insisted that his brother Alex, who had a similar business at Strone Point, Argyll, should be a bigger customer, but we never managed to sell him anything. Perhaps because he intended to retire, to Mexico of all places. He did sell the business, to Robin Gibson, made redundant by Bernat Klein at Gibson and Lumgair, and we did develop an ongoing trade with him. Alex Macintyre still lives in Kilmun!

The origins of the Edinburgh Woollen Mill are worth recounting. There was, in Langholm, a very successful package dyeing company, the Langholm Dyeing and Finishing Co. Ltd. The brain child of and managed by Andrew Stevenson and financed by him and the Bell family of Arthur Bell and Co., also of Langholm. Andrew had two sons, the elder being David who, in 1970, represented Scotland in the pole vault event in the Commonwealth Games in Edinburgh.

Two things set David on the trail which was to make him the biggest and most successful of the tourist shop operators. The first was those games. He thought there might be a market among the competitors for tartan cloth to take home, so he had Edwin Armitage at the Woolly Mill in Langholm make a few half pieces of tartan which he peddled in the competitors village in the evenings. Successfully. The second event concerned a dyeing error by Langholm Dyeing, where he was by now working. A dyeing by them for Smith and Calverly in Yorkshire was off shade, and even though it represented only a small part of the design a hundred or more pieces of cloth could not be delivered to Smith & Calverly's customer. Naturally they laid the responsibility at Langholm Dyeing's door, and young David was sent down to sort it out. He failed to agree a figure for a claim based on the price Smith & Calverly thought they might get for the cloth jobbed off as seconds, so he hired a lorry, bought it from them, and brought it home. He sold it very well around the mill and tourist shops, including one which he opened for the purpose, and, largely as a result of that experience decided to found a chain of shops.

The venture prospered beyond his wildest dreams and grew to be a major customer for the cloth, yarn and knitwear industries of Scotland and Yorkshire. Gardiners's association with Edinburgh Woollen Mill (EWM) began by selling small quantities from our standard range, moved on to making substantial quantities of woollen tartans for them and, when that branch of their operation "went worsted" changed to weaving Shetland yarn provided by EWM, and then to selling them yarn from the new spinning mill, a large part of which came back for weaving. At it's peak this business was worth three thousand pieces a year and, as with Moffat, negotiation produced a tough price but back end of the year weaving to plug the spring hole. The business went on until early in the eighties, EWM bought Heather Mills, a weaver in Selkirk, which made a considerable mess of Gardiner's spring season and was instrumental in provoking a weaving reorganisation of which more later.

For a number of years Munrospun, of Edinburgh, selling traditional tweeds to the tourist trade and to the British retail in general under their famous trademark Munrospun, "wool's name for luxury", had been a significant customer, but they now entered a period of terminal decline. A Selkirk firm, George Rae and Sons Ltd., run by Mary Rae, picked up a lot of the pieces and became a major customer and ultimately a major problem. Another large Scottish client was Richardson Tweeds, of Galashiels, where Bobby Richardson had developed an enormous mail order trade with school sewing classes, and used Gardiner cloth by the hundred pieces to supply it. Again the trade lasted many years until a combination of mistaken design selection and a month long postal strike brought Richardson down, leaving Gardiner with it's first major bad debt.

And, at the end of the period, Gardiner appointed Alf Masterman agent for menswear fabrics in London, which brought Burberry's account into the fold, an account which was to grow in importance over the years until, as all accounts do, it "went away" in 1989.

When the accounts for the year to January 1973 were approved the conflicts, as they affected Gardiner, between the various legal restraints made themselves felt, for dividend limitation allowed an increase from 5% to 5.25%, but wage limitation would not have let the Executive Directors, who waived their dividends in favour of matching bonuses, have that increase, so the dividend remained at 5% though, had the shareholders agreement on dividends been allowed to work, the rate would have been 20%. With the higher tax rates at the time above 80% Colin and I had seriously considered suggesting that instead of bonuses we had a Rolls Royce apiece! But dividend and wage limitation scotched that idea.

The limitation of wages led us into dispute with our shop stewards about payment for tuners for working the four extra machines delivered in December 1972. The law said that nothing could be paid but the tuners were unhappy about that. After much argument and half promises about what might happen when freedom returned they agreed to work the machines, but as a token of resistance to government policy they persuaded us to break the Government's phase 3 limit of an increase of £4 and pay £4.01! Wage restraint undoubtedly soured discussions for the whole of the period in which it applied, for with ever changing conditions vital adjustments should have been made, but were forbidden. Naturally we, management, were accused of hiding behind the legislation but as penalties for breach included the repayment of capital grants which by then exceeded £100,000 we felt we had to tread a little warily. Not only were wages and dividends restrained, but so were profits and in one year Gardiner exceeded the allowed increase. Fortunately, before the expiry of the year in which the compensating correction should have come through the restraints had gone by the board. None of this made conducting the business any easier and indeed, in March 1974, with the new and expanding printed circuit industry in the Borders soaking up more and more of the better quality labour available we, entirely illegally, lifted all wages by £2 per week and all shift allowances by £3. The annual cost of this was £45,000, but something had to be done to stop the drain of labour. Again, the rules changed before we were found out.

To add to all the problems of restraint the wool market was in one of it's periodical frenzied spells. The first half of the January 1974 year produced a profit of £96,831 after providing £87,000 for stock and contract losses, "which matches the bonus profit made in the half year from the consumption of cheap yarn contracts". That swing became a roundabout in the second half, for the provision was substantially absorbed by falls in wool prices so that the second half yielded £86,375 for a record year's total of £183,206. And that should have been more, for in the back end of 1973 we were contemplating a profit of a quarter of a million, but a spring season failure coupled with production difficulties in orders for black and white herringbones, a most difficult design to make in the absence of perfectly level yarn, saw our dreams of glory drift away. Even so that result was probably, taking inflation into account, as good as 1988's record £579,256 from spinning and weaving together.

Earning nearly thirty percent on capital Gardiner was comfortably outpacing the English Calico standard of twenty. West Germany had suffered a setback at the end of 1973, but there was no reason to believe that it would not recover it's growth pattern. Everything looked good for our corner of the weaving business, but our experience of the roller coaster nature of the trade, of the inevitability of cycles in it and of the vagaries of fashion made us reluctant to expand the weaving operation beyond the five hundred pieces a week level which was the current target. The last six Sulzers should keep that target in range with any foreseeable movement in cloth weights, Gardiner had reached it's maximum development as a weaving business.

Chapter Seventeen

The Grass is Greener? 1974

The options open to a weaving business which is believed to have reached the limit of it's organic development are limited to going upstream or downstream. Downstream expansion would mean going into garment making, an option which Colin and I rejected for a number of reasons the principal two of which were, firstly, that by so doing we would enter into competition with our existing garment making customers and, secondly, that there would always be the temptation to use Gardiner fabrics when they were not really suitable, or to force Gardiner into making things which it should really leave alone.

The upstream option was to build a spinning mill to displace the existing spinners and to supply the main yarn qualities used by Gardiner. For a number of reasons this was an altogether more attractive idea. At the crude level of profitability it would transfer the profits currently earned by our spinners from supplying us into our own pocket. Secondly, it seemed probable before we did the arithmetic that we could make yarn, on modern machinery, more economically than it was being made at present. And, thirdly, it ought to be possible to make better yarn on a thoroughly modern plant.

So, after weeks of morning mail time chats we decided to explore the mechanics and financial consequences of building a spinning mill big enough to supply our basic needs, and it went without saying that the utmost secrecy must be maintained throughout the exploratory process, which limited our investigative field. Clearly what we needed was a consultation with someone who could supply virtually a complete plant and who would be able to calculate the machinery needs and price the whole exercise. That man was close at hand.

Harvey Wilson had been running The Border Textile and Engineering Supplies Co. for a number of years in succession to his father. Originally set up to supply "mill furnishings" to the trade it had expanded first into selling Hattersley looms, and the warp mills also made by that firm and then, in the early seventies, had got the Scottish agency for Hoeget Duesberg Bosson carding, ring spinning and twisting machinery. HDB were a Belgian firm, one of the big three in their field, and Harvey got the agency in time for the massive re-equipment of the spinning industry which Regional Development and Industry Act Section 8 grants brought about. So we sent for Harvey, swore him to secrecy and outlined to him our needs in terms of weekly weight of the two basic qualities, known to us as C16 and S24. Harvey went away and a few weeks later handed over to us the plant specification of the machinery required to do the job.

His proposals called for a kit of HDB machinery comprising a complete blending and willeying plant, a baling machine, three 2.5 metre wide carding sets and eight 160 spindle frames. The basic cost of this was set at £663,372 and a great deal of further equipment was needed to support this productive unit. Automatic winding machinery to package the product, conditioning equipment, vacuum fettling gear and grinding machinery for the cards, the initial clothing of the machines in card wire, fork lift trucks, barrows, bobbins and a host of other small equipment had to be priced and added into the capital programme. And, most expensive of all, a 3,200 square meter building, air conditioned and with standing headroom pits under the carding machines had to be designed, priced, and added to the "money needed" figure. With all the capital costs known estimates of working capital requirements were made, labour and output budgets prepared and yarn costings derived from them, and at that point we nearly, by reason of our imperfect knowledge of the spinning process, gave up. Our costings came out more expensive than those of either Pogson or Shires. For weeks we agonised about it until, who knows how, we discovered and became masters of "the oil argument", a very simple concept once you have discovered it, but, as we were later to discover, one which otherwise brilliant men in the hosiery industry could do their best to say they didn't understand.

The problem is this. Yarn is priced per kilo. The length of yarn per kilo depends on the "count". In Galashiels notation the number of

200 yard lengths in a pound is the count. Thus 16 cut, our C16 yarn, would contain 3200 yards per pound. The yarn spun in Yorkshire was made with an oil content of 12%.

The yarn we proposed to make on fast new machinery would have an oil content of only 3%, so if you made 3200 yards of that to the same CLEAN count it would not weigh a pound. And in cloth construction it is clean count that matters. Therefore, the clean count of the yarn we had been buying was 100/88 of 16 cut, or 18.18 cut, and if we made yarn to the same clean count we would have to spin, in 3% of oil, to 17.63 cut.

In other words direct comparisons of greasy prices per kilo only work if the oil content is the same. Only by reducing everything to clean counts can true comparisons be made. When we tumbled to this and factored it all into our calculations the picture looked much more promising. It was not until we were about to put our proposals to the Board, with a built in price disadvantage, that the light dawned.

The next task was to see the machinery that Harvey was talking about in operation and that raised problems of secrecy. We couldn't visit any of his United Kingdom installations for, the grapevine being what it is, our spinners would have been informed within minutes, so we embarked on a series of extremely enjoyable outings to visit plants abroad. First, in March 1974, Harvey and Colin and I went off to Denmark to visit a two set mill in Holstebro in Jutland. Even getting there had it's moments. We flew from Newcastle, there was fog at Heathrow, and our 'plane was stuck in London. So Colin and Harvey, experienced travellers both, got out their timetables, searched the departure board and sought ways of getting indirectly to Copenhagen, by routes as diverse as Amsterdam and Warsaw, Milan and Dusseldorf or any other unlikely combination of obscure waypoints they could cook up. It took the stay-at-home me to notice the flight to Kristianstadt, just across the sound from Copenhagen, but happily our own flight turned up, not too late, and we didn't have to do a tour of Europe and Asia.

The mill in Holstebro was a lovely, clean little place making furnishing yarns. The Proprietor had an enormous Newfoundland Hound as soft and friendly as a little lamb in his presence, but, he said, murder without him, which was why he left it guarding the office while we were in the mill. Lunch followed. Genuine Danish Smorresbrod. Course followed course of pickled fish washed down with aquavit and lager, until we were replete. Then the steaks appeared. How we survived without bursting I know not. It was quite one of the best meals ever. Back to Copenhagen for the night, then home on the morning flight to Newcastle. The next expedition was to see two mills spinning wool yarns in Castres in the Department of Tarn. Flying into Toulouse we saw the Concorde prototype on the runway. Once again eating and drinking enlivened the trip, with the hotel in which we stayed producing the most excellent dinner. The first mill visited was a small family outfit where at lunch time everything stopped, trestle tables and tablecloths appeared and a civilised, wine washed lunch was consumed by the workforce before production resumed. The principal was a rugby fanatic - apart from the players from Racing Club de Paris most of the French team come from this small corner of France. The principal and his pals from the Castres Club were reputed to have dug up the Champs Elysee on one of their visits to the Capital. The next visit was to a mill delightfully situated by a swift flowing river, with the owner's house perched on a cliff overlooking a gorge where we were fed Pernod till we looked yellow. It was here that we first saw the automatic winder that we eventually bought, made by Gilbos, of Belgium.

Home again, and we engaged the services of one Robert Rae, lately Managing Director of the Dawson Group's Waverely Mill in Innerleithen, to consult with and advise us. Robert had vast experience of spinning, had lost an arm in the war and still played golf to a 12 handicap. His services were invaluable to us, his most noticeable contribution being that he made us think seriously about Tatham carding machinery, made in Rochdale. Colin had looked at it, without giving the appearance of any special interest, at one of the other Dawson mills which he was visiting on National Association business, but had not concluded that it was better than HDB. Robert Rae insisted that we have a proper look and one fine day the three of us, plus David Porrit, of Tathams, flew off from Manchester bound for Dublin and thereafter by road to Cork to visit a large carpet spinning plant belonging to Youghal Carpets. More superb eating and drinking but more importantly we were sufficiently impressed to seek a quotation for three of Tathams 2.5 metre cards.

By now Colin and I were thoroughly hooked on the idea of building a spinning mill. We really believed it to be necessary if Gardiner were to continue going forward. We gathered all the facts, figures and opinions together into a paper on the subject and presented it to the Board for discussion at a meeting on 12 June 1974. The full Board, Colin, me, Bob Audsley and Norman Hornsby were present. By now they were representing Tootals, the new name for what had been English Calico, a name taken from the tie and handkerchief subsidiary Tootal Broadhurst Lee. The subject was thoroughly aired and the meeting concluded:-

" It was agreed that a spinning mill be built as suggested subject only to the following conditions being fulfilled.

1. That the necessary loans could in fact be contracted for with the Royal Bank of Scotland Ltd. The Industrial & Commercial Finance Corp. Ltd. and/or the European Investment Bank.

2. That sufficiently firm arrangements could be made for the provision of housing for incoming workers to man the project."

It was noted that a new, later, schedule incorporated a more correct calculation of the cost of production of our 16C Cheviot yarn and that this was more favourable to the project than the original calculation. It was also noted that should everything go exactly to plan a spinning profit of between £200,000 and £230,000 would be achievable which, allied to possible savings and increased efficiency detailed in another schedule would, on the basis of the most favourable of the new cost calculations, result in a return on capital of acceptable levels without the need to increase cloth prices.

It was agreed that before the orders for the machinery were placed Mr. Wagstaff (Tootal's Finance Director) should have an opportunity of going over the figures and that it would be advisable for him to visit Selkirk for that purpose.

Irony. The day that our plan became public knowledge, Gibson and Lumgair, one of our competitors in Selkirk, announced that they were to close their spinning division for, by so doing, they would gain flexibility!

Chapter Eighteen

The Spinning Mill 1974/1976

With the project agreed in principle the next, and very urgent task was to find the money to finance it. Our first visit was to the Royal Bank where Robert Cumming, the Southern Region General Manager, agreed to cover part of the working capital requirement by increasing the overdraft limit to £200,000. In the course of the discussion he asked whether Tootal would guarantee the overdraft but accepted our position when we said that we would not ask them to guarantee anything. The figure of £200,000 was an interim figure with the Bank committing themselves to us to go further when the need for the money arose and in the event March 1976 saw the limit raised to £420.000, and then again in April 1977 to £500,000. We were well supported by the Royal Bank, even when, in 1976, the overdraft diverged from the forecast. When things went a little awry they insisted on a separate wages and salaries account to increase their security, and sought, and got, monthly long range cash flow forecasts but never at any time did they suggest pulling the rug from under us.

We visited ICFC in Edinburgh, where sharp young men with little understanding of manufacturing industry and no stomach for risk turned us down. As did the European Investment Bank. The sheer size of the project relative to the existing capital base deterred lenders and, in September 1974, a Board minute reads; "It was noted that efforts to obtain finance, other than from Houget Duesberg Bosson, for the carding machinery, had so far failed, chiefly on the grounds that the Company would be embarking on too great a degree of indebtedness. It was noted that this view was not without support in the higher management of Tootals."

The relating of the borrowing problem to the carding machinery was because we had, by that time, decided to use Tatham cards, if we could finance them and while HDB, who were Belgian Government supported, would finance their machinery, Tathams, who were private, would not.

By that month of September it began to look as though the first condition, finding the money, would not be fulfilled and gloom prevailed at Selkirk, though the prospect of failure to find funds was beginning to be contemplated with favour in Manchester where a new Chairman, Michael Kirsopp, had replaced Sir Neville Butterworth. But it is always darkest before the dawn, and Selkirk Town Council, anxious that the project go ahead, offered to finance the new building. Hope turned to despair when we discovered that they were only allowed to lend against buildings on their own land, then back again to hope when, at a meeting in the Council offices with Councillor Len Thomson and Town Clerk Robertson they decided that the fact that they owned the feu constituted ownership. It looked to us like stretching things, as the feu had about nine hundred years to run at a duty of something like £14 a year, but who were we to complain? A loan of £225,000 to cover 75% of the building costs, at fifteen and one eighth percent, repayable over thirty years, was arranged.

Between June and September we had a visit from Mr. Brown of the Department of Trade and Industry, who spoke at length on the Industry Act 1972, and particularly of the grants available under Section 8. This section dealt with the wool textile trade and provided grants for the closure of mills, and against new plant and buildings where old were scrapped, on a replacement of capacity basis. But we had no spinning plant to scrap, and the discussion, though interesting, was going nowhere until we asked Mr. Brown whether, if we bought plant from Pogson, left it in place and leased it back to him to operate in the interim, scrapping it when we no longer needed it's output, that would qualify us for grants. Mr. Brown went away to think about it and ultimately said yes.

Though we didn't realise it at the time we had invented the transferable scrapping certificate which, in its ultimate refined form, allowed the simple transfer of the certificates which entitled a firm to· grant without the need for the sale and lease back interim state. In the event we did agree to buy plant from Pogson, we did lease it back to him for nominal rent and it was ultimately scrapped on our behalf. And we did get the grants. But that is running ahead for, if secrecy were to be maintained, we had to be able to finance the project without Section 8 grants. By now the second condition, housing for incoming workers, had faded in importance, for recession was upon us, and recruitment looked like being a minor problem.

By the end of September the last of the finance was in place. Tathams had decided to lend us the money to buy their cards over five years and Prouvost & Lefvebre, our principal wool merchants to be, had agreed to plug the £75,000 gap which remained to be filled, the gap being reduced by the decision to use Tatham cards costing £50,000 less, each, than the HDB machines. The final brick in the wall was the decision to scale the plant down to two carding sets and five frames to fill two thirds of our need, whilst keeping the building at three set size. We were ready to place the orders, and did so at the beginning of October.

Once the orders were placed secrecy was a thing of the past. It was vital that our spinners, Pogson and Shires, upon whom we would have to rely whilst the spinning mill was building, and thereafter for a third of our supplies, heard the news from us before it came to them from other sources and I took on the job of telling them. I visited Gerald Shires and gave him the news in his office. Gardiner was by no means as important to Shires as it was to Pogson, so I arranged to have dinner with Raymond and Neil Pogson at the George in Huddersfield. I decided to make a party of it and engage a private room, for which there was a precedent as I had given dinner to them and Joe Nutton, their wool merchant, on the occasion of our centenary, also in a private room.

The day arrived, I drove down to Huddersfield and when the brothers Pogson joined me I said that, before we got down to dinner I had to tell them that we were going to build our own spinning mill. The response astonished me, and underlines my belief that the Textile Industry resembles nothing more than a large scale ladies tea party.
"Aye, lad, we know" said Raymond,
"You can't know, we only ordered the machinery today", said I, "How do you know?"
"Well", said Raymond, "We didn't know, but we were pretty certain.

Have you been using Robert Rae, with one arm and a 3.4 Jaguar, as consultant? Because he sought a quotation for clothing three 2.5 metre sets, for an undisclosed client, from Chris Tetlow, the card clothing manufacturer, and yesterday, driving past your offices, Chris saw Robert Rae's car parked in the yard and Robert himself in Colin's office, through the windows. He put two and two together and, knowing how much we did with you, raced down to Slaithwaite and told us".

Our friendship was an enduring one and survived this shock. We had a very good evening and usefully negotiated the sale, for £30,000, of enough of Pogson's oldest machinery to Gardiner to get us our Section 8 grants. Upon which, to our horror, the Inspector of Taxes would allow neither annual nor, on scrapping, balancing allowances.

1974 was one of the years when the British economy really fell to bits, culminating in the Burmah Oil fiasco and the bottoming of the FT index at the turn of the year. Many a respected industrial share was standing below par. The Royal Bank was one, and Tootal was another, and whilst Bob and Norman appeared confident in the spinning project others at Tootal were getting increasingly unhappy. Our cash flow projections already showed that we would be on the ragged edge if spinning did not become profitable quite early in it's life, and Tootal themselves were looking at cash problems. As early as August 1974, just two months after giving approval to the project, the Tootal cold feet prompted Audsley and Hornsby to call for a meeting. In the event only Hornsby and I met, at the George in Penrith, and it soon became plain that Wagstaff and Wetherby wanted to use their disquiet as a means to get Tootal out of Gardiner altogether. Apart from the usual complaint that Tootal got nothing out of Gardiner a new one was added, to the effect that the project would crystallise that situation for years to come; "so when do we get our reward?" This was the easily identifiable forerunner to talk of share transactions which this time took the form of a suggestion that Colin and I sold our part to Tootal.

A long, rambling discussion of price followed, with a figure produced in 1973 by Bob Audsley of £187,500 per corner as a base, and used by me to support a suggestion of £200,000 per corner. Norman countered with £125,000, derived, I think, from the recent purchase by Tootal of a net curtain factory making £100,000 a year for £400,000. This proposition then died and we moved on to talking about a joint sale in five to seven years, when we would, should we prosper, have an asset value of one and a half million, the meeting drifted to no conclusion after a discussion of the uninviting prospects for the coming year.

Not very many weeks later Colin and I were summoned to Manchester to talk about the possibility of our buying Tootal out, a proposition unlikely to attract us at any price they might think right. Our experience of earlier circuits of this track was repeated yet again and the meeting did not last terribly long. With the Tootal share price standing below par a little arithmetic with p/e ratios made it clear that the sort of price Kirsopp and his colleagues had in mind was not in line with the market, even ruling out what they clearly regarded as the great risk of the spinning project, now contracted for. No progress was made and relations between us and Tootal slipped a little further. We went home even more determined that the spinning mill would be a success despite their fears.

Early in January 1975 the subject was raised yet again, and we were asked to go to Manchester for another discussion. We refused to go, being sure that it would be just another trip round the same old track but Tootal wouldn't be denied and we finally agreed to meet Bob and Norman at the George at Penrith, "to talk about B & J buying the Tootal half. Before we set off I scribbled on a piece of paper the terms on which I thought we might buy, and agreed them with Colin. They were: -

-     Tootal lend us the money.
-     Free of interest.
-     Repayable only when we sell Gardiner.
-     And then only to the extent that the sale price covers the price paid.

We met in a private room for one of the roughest meetings of our lives. Bob and Norman were clearly under huge pressure from Kirsopp, Wagstaff and Wetherby to get Tootal out of Gardiner, and the basic proposition was that either we bought the shares or we cancelled the spinning project, which had by now reached the stage where the building steelwork was up and HDB and Tathams had started making the machinery. And our spinners were seeking other markets. Cancellation would have brought huge penalties upon us but the response to "will Tootal pay them for us if we cancel?" was "NO". When we pointed out that Gardiner paying would bankrupt Gardiner it cut no ice at all.

The next move was for them to suggest that they would advise all our creditors that they had disowned us. Whilst technically this would be an ineffective course of action in practice it would shatter confidence in Gardiner both up and down the line with dire consequences. Our response to that was that if anything like that happened the Editor of the Daily Express would surely hear of it. Round and round we went, with me and Colin simply refusing to consider buying, for we really didn't want to buy nor to provoke a fruitless price argument, and Bob and Norman insisting that we bought or cancelled.

After a couple of hours it was plain to all that no progress had been made, nor was any likely, and Bob Audsley broke the deadlock. Very quietly he said "YOU WILL BUY OUR SHARES. We will lend you the money, free of interest, and you needn't repay us till you sell those shares". We were gobsmacked, and after a short, quiet chat we agreed, I think we said that in a deal like that price was irrelevant, but a price was fixed, though there is no record of what it was. There being nothing left to talk about we went below for a convivial lunch, and set out for home in our opposite directions.

Immediately on reaching Selkirk I wrote to Tootals setting out the agreement we had reached. Two weeks went by with no acknowledgement but then a letter from Mr. A Tyldesley, the Group Secretary, arrived.

" Mr. Hornsby has passed on to me your letter of 15 January. The matter has been discussed by members of my Board and it is clear that the terms outlined in that letter for the acquisition of our shareholding are so harsh as to be incapable of acceptance by us."

The letter went on to offer the holding to us for £120,000 or, alternatively proposed a reconstruction which left us with all the equity and Tootal with £240,000 of debenture stock redeemable in seven years, both propositions which we, with Gardiner embarked on the risky spinning project and recession upon us, could not accept. The real sting was in the last paragraph.

" If you are not prepared to agree to accept our original offer or an arrangement on the above lines within the next 28 days, then we shall feel free to make it clear in writing to any organisation that has a substantial financial involvement with the Company that Tootal Ltd. has not provided and will not provide any finance or guarantees of financial support in any form whatsoever in respect of Edward Gardiner & Sons Ltd. whether for the purpose of financing expenditure on fixed assets or of providing working capital."

We waited 28 days. Nothing happened. Except that Alan Wagstaff, the Finance Director, rang to say that he wanted to meet us for dinner in Edinburgh one evening, and fixed a date.

We arrived at the Caledonian Hotel one foggy evening to find that Alan's plane was late, and that he had taken a suite, bar, lounge and dining room, for our meeting. He arrived, full of apologies for his fog induced tardiness, and announced immediately that he was here to make us buy the Tootal holding. We said, equally bluntly, that we wouldn't. "OK then, that is that. Let's enjoy our dinner. You can forget the bit about telling all your creditors, but I feel I must tell the Bank of the state of the game".

Alan is a very nice man and we had an excellent dinner and evening, and agreed that Alan should come with us when we went for our annual meeting with the Bank in May to apprise them of the Tootal position.

Within a fortnight both Bob Audsley and Norman Hornsby resigned from the Gardiner Board, upon which Tootal was never again represented.

In May, true to his word, Alan came up and visited the mill, and we went on in the afternoon to see Robert Cumming at the Bank's head office in St. Andrew's Square. After the usual discussion of the weaving trade, the state of Gardiner's finances and general position, and a report on the progress of the spinning project, we agreed the overdraft limit for the coming year up until the spinning mill began to increase it, when more talk would be in order. Robert then turned to me and said that that concluded our business, and thanked us for coming in, to which I replied that that was not quite so, for Alan Wagstaff had something to say to him. Whereupon Alan said his piece, that Tootal would not prop Gardiner and had withdrawn from the Board to conserve management (Tootal's) time and effort. It was brief and to the point.

Robert leaned back in his tilting, swivelling leather armchair behind his crescent shaped desk, put his fingertips together under his chin and responded as follows:

" Thank you Mr. Wagstaff, I find your Company's decision singularly irresponsible, but it is your decision to make. I do thank you for taking the trouble to come and tell me."

We took our leave. Contact with Tootal was maintained over the next thirteen years through Norman Hornsby, who had connections in the Borders and visited us once a year or so. The Gardiner articles specified not less than three directors, so we appointed our wives to make up the numbers.

Bob Audsley went on to the Chairmanship of Tootal after Kirsopp's brief reign and sadly died in office. Norman Hornsby soldiered on as Director of the Clothing division until he retired, leaving us with the sombre message "when I've gone there will not only be no-one at Tootal that knows about Gardiner, there won't even be anyone that knows where it is."

The causes of this final break with Tootal lay, firstly, in the constant ploughing back of profits at the expense of dividends, though a perfectly satisfactory dividend agreement had been made for implementation when the law allowed. Then secondly in the sheer scale and risk of the spinning mill project by comparison with the weaving base. Thirdly, we believe that Bob Audsley quite rightly in theory but perhaps wrongly politically regarded the project as a Gardiner matter and did not fully inform his colleagues. And lastly, the onset of recession and Tootal's own cash problems caused them to look at everything with a jaundiced eye.

This spinning project, which was to treble the fixed assets of the Company and increase it's net current assets by more than fifty percent was to be built on a capital base of £300,000 of equity and £6000 of Preference Capital. The existing articles allowed borrowing up to one and a half times the issued capital. The solution to the problem involved both amending the articles to allow the Directors to borrow up to three times the issued capital, and an increase in the issued capital itself by the capitalising of reserves, £16,000 from the Capital Redemption Reserve and £84,000 from the Profit and Loss Account, to make the Ordinary Capital £400,000. To tidy things up the last of the Preference Shares were redeemed on 31 March 1975, which was too late to prevent all the resolutions involving the increases of capital and borrowing powers proposed at the Extraordinary General Meeting on 24 March 1975 being rejected because, from the minute;

" It was noted that every resolution proposed would, if passed, be invalid by reason of the existence in issue of £4,000 of Preference Shares between this date and the date set for their redemption, 31 March 1975.

The resolutions set out below are therefore all rejected".

Black mark to the Secretary.

The meeting was reconvened for 25 April, by which time there were no longer any Preference Shares in issue, when all the resolutions were passed. Gardiner could now, legally, undertake the spinning mill project on the basis of an Ordinary Share Capital of £400,000, half owned by Tootal and a quarter each by Ian Jackson and Colin Brown, no Preference Shares and a Director's borrowing limit of £1,200,000.

The rest of 1975 was spent watching the spinning mill building being put up and fitted out with it's services, air conditioning, electric power, water and the like and then overseeing the installation of all the machinery, the delivery of the initial wool stocks and the appointment of the key members of the staff. David Knox came down from Johnstons of Elgin to manage the spinning operation, Hugh McGill joined to run the testing laboratory and Colin Ford was persuaded to leave Ettrick and Yarrow Spinners, next door, to be our "shader", the man responsible for getting the components dyed and put together to make the correct shade. The plant was to run fifteen shifts a week, so three carding engineers were needed, and these, Sandy Allinson, John Shiell and Michael Stewart were all appointed in time to see the machinery installed, as were the Head Spinner, Jackson Cockburn and the Wool Store Foreman Jim Muir. The machine operators were engaged later, some of them coming from the weaving division's twisting department which was being reorganised.

As the bills came in the grant claim forms were completed as swiftly as possible, though the two major bills, apart from the building, would only come in in the form of instalments repaying the loans from the makers, but the ancillary equipment accounted for a tidy sum and the overdraft mounted. On remeasurement the building was found to have come in substantially under budget and Jimmy Beck, on behalf of Kelvin, recognised this by allowing Gardiner a very substantial rebate from the contract price.

As December approached excitement and tension mounted as wool was put through the machinery for the first time and the first actual production batch was ordered by the weaving division from the spinning division. This was a batch of shade 825 grey in our quality C16 (140 tex cheviot), ordered on 14 December 1975 for delivery on 19 December, signed by George Lindores. It was made and delivered on time (perhaps we wrote the order retrospectively), and the long struggle to turn the beautiful new mill into a productive and profitable plant began.

Meanwhile, back in the weaving division a good first half to July 1975, £87,738, was followed by the second half loss of £35,804, giving a year's profit to January 1975 of £51,934. The second half had been severely hit by recession, short time working having started as early as September. It went through until mid April 1975, seven straight months of week about or it's equivalent, the longest spell in Gardiner's history.

In May 1975 Robert Rae joined the Board to guide us into the spinning business. The appointment was for a two year stint, but Robert remained on the Board until the end of 1979.

The downturn in trade prompted the usual heart searchings about product and product policy and Gardiner's current success in the tartan market generated the idea of expansion in that field, buying four more six colour Sulzers and selling four of the original tappet machines to help pay for them. With grants and the proceeds of the four tappets the likely cost was less than £20,000 but the transaction never came about. Delayed by the commercial gloom the idea was overtaken by events.

Furnishing fabrics were one of the avenues explored, for they tended to be counter seasonal. First of all Chilton, Phillis and Guest tried to indicate and sell fabrics, and when orders failed to materialise expert agents, the Brian Weyland organisation, were appointed. They, too, failed.

The year to 31 January 1976 closed with a loss of £8,076, the first loss for 14 years. The first half had made £60,665, but the second half had lost it all, though £19,500 of the loss was a single bad debt, which was recovered the following year.

The account concerned was George Rae and Sons, who had followed on the Scottish Country Tweed theme as Munrospun declined. Over the years our business with them had grown but their payments to us had declined until they owed us more than £100,000, a position which we should not have allowed ourselves to get into. Somewhere along the line we had obtained a charge on their assets, and in 1975 we went the length of putting in a receiver, Frank Mycroft, of Deloitte, Haskins and Sells. First of all Frank unwound the deal which I had set up with David Stevenson at the Edinburgh Woollen Mill for him to take over the stock, then he turned the business round, then he handed it back to Mary Rae as a going concern, having paid us the whole of our debt. A most impressive performance. So had we been able to see, and take account of the future, the year to January 1976 would have shown a profit, not a loss.

Weaving division gloom cast it's shadow on the spinning mill. With the former on short time the party which we had planned for the latter's opening had to be abandoned and the new venture crept into life with no bang at all. But despite all the strife it was there, and working.

165

Chapter Nineteen

Spinning: The Early Years 1976/1979

Management's conduct of Gardiner's affairs over the next few years has to be set against the background of changing conditions in the market for woven woollen cloth and in particular the increasing severity of the spring season problem. The figures for the calendar years 1976 to 1987, in Gardiner's terms from the year ending 31 January 1977, the first year of the spinning mill's operation, to the year ending 31 January 1988, set out below, almost tell the story without words.

Year To 31/01	First Half			Second Half			Full Year		
	Spin	Weave	Total	Spin	Weave	Total	Spin	Weave	Total
77	(58742)	94052	35310	16495	1549	18044	(42247)	95601	53354
78	80747	27378	108125	53758	18787	72545	134505	46165	180670
79	85908	116512	202240	20782	(45458)	(24678)	106690	71054	177744
80	65934	32299	98233	86058	(97288)	(11230)	151992	(64989)	87003
81	25674	109751	135425	46373	(115423)	(69050)	72047	(5672)	66375
82	105436	123594	229030	37433	(30840)	6593	142869	92754	235623
83	7294	196021	203315	(59351)	(198383)	(257737)	(52057)	2362	(54419)
84	4906	58681	63587	109413	(82624)	26789	114319	(23943)	90376
85	13328	83096	96424	67054	(88856)	(21802)	80382	(5760)	74622
86	43770	133913	177683	130799	(130843)	(44)	174569	3070	177639
87	98557	103183	201740	6957	(49892)	(42935)	105514	53291	158805
88	175000	296000	471000	132000	(23744)	108256	307000	272256	579256

The experience of starting up the spinning mill was quite unlike that of moving from conventional looms to Sulzer weaving machines where we were merely transferring an existing product to a new type of machinery. This start up was of a trade wholly new to us with Colin's Laidlaw and Fairgrieve experience many years in the past. Very early in 1976 Colin and Robert Rae set up a regular Monday meeting with David Knox to review progress and initiate steps to move production towards the targets set at the feasibility study stage, no simple matter for two and a half metre cards were not common in the industry and none of our people had experience of working them.

Part of the preparation for the start up of the spinning mill had been the commissioning of ICL's programme writing subsidiary to provide a suite of programmes to deal with spinning operations so, from the beginning, proper figures were available to the Monday meetings for yields, efficiencies, dyeing costs and the like. Week by week the Monday meeting wrestled with the task of kicking the new venture into shape, a matter made the more urgent by the difficulties in the weaving side. Recession and the spring season problem were making that trade less profitable than hitherto, with the achievement of any profit at all in the second half of the year becoming more and more of an uphill task. The second half weaving profit achieved in the year to January 1978 was the last in this history. It was the hope of all of us that spinning could provide year round profit to damp down the seasonality of the weaving result.

Throughout the first half of 1976/77 the struggle to achieve target production went on, and often looked like a losing battle. Adjustments were made to the wool blends used for the basic qualities which, though pushing up the fibre price, brought efficiency gains which paid for those increases, but by the end of the half year to July, which showed a spinning loss of £58,742, all the efforts of the Monday meeting began to be rewarded. As the second half developed production crept up towards the target and early in the period the monthly result moved from loss to profit, with the new mill's second half contributing £16,495 to the overall result. Spinning, in the next eleven years, never failed to produce a first half profit and only once, in the depths of the recession in the early eighties, did it show a second half loss.

The ups and downs of the two major mergers in the Scottish weaving industry had resulted by 1976 in the shutting down of all the component mills but those of David Ballantyne, in Peebles, and Henry Ballantyne, now part of the same Group, Scottish Worsteds and Woollens Ltd., under the leadership of Henry Ballantyne, of the D. Ballantyne Company. Members of the Ballantyne clan were to influence Gardiner's progress through two different contacts and two entirely different activities. The first involvement was with Henry Ballantyne and was so short as to be almost unnoticeable. It was as a Director of D. Ballantyne that Norman Hornsby gained his Scottish Woollen experience and in the course of Tootal's efforts to extricate themselves from the spinning mill project he was prevailed upon to suggest to Henry Ballantyne that his group took over Gardiner.

Henry visited us, was shown the whole operation, including the then building spinning mill, and came up with the suggestion that he would consider taking over provided that the weaving division was closed down! Though in later years shutting that division was seriously considered it was not then on the agenda, and Colin and I refused to entertain the idea. So Henry went away.

The Managing Director of the Henry Ballantyne mill in Walkerburn in the early seventies was Jeremy Ballantyne. As the apparel side of the weaving trade became more difficult he recognised the opportunities in the field of fabrics for curtains, office partitioning and the like. The group already had a successful toehold in the upholstery fabrics field through their subsidiary British Replin, of Ayr, but Jeremy failed to persuade his Board to back him in curtain fabrics, resigned, and set up his own firm. He leased a new factory in the recent development at Tweedbank in Galashiels, bought suitable weaving machinery and set off to build himself a business. All was going well, initial losses had been turned into profits and he was beginning to become recognised and established in his new trade when an arsonist struck, burning his plant to the ground in May 1976.

Jeremy immediately set about finding commission weaving for his product until he could rebuild and Gardiner undertook some if it. Very quickly Jeremy realised that rebuilding would not only be a hearbreaking task, a real struggle if he had to run the business on a commission basis at the same time, but would also require, insurance notwithstanding, an injection of capital, and paid us the compliment of coming to talk to us about his problems.

The upshot of these talks was that he decided not to rebuild. Gardiner agreed to buy two wide and two standard Somet looms (ideal for making the Ballantyne product), and to enter into a five year contract to manufacture all of Jeremy's fabrics. Starting prices were agreed and an elaborate formula incorporating the Retail Price Index and the index of Textile wages was devised to keep prices in line with current conditions without the need for frequent arguments. Gardiners owned the stock, and the Work in Progress, invoicing the finished product to Ballantyne when it was despatched. Jeremy himself was concerned with designing and selling the product. Gardiners agreed to cease their own very small furnishing business whilst retaining the right to trade in upholstery fabrics, a right which never yielded Gardiner any benefit for we failed to make a mark in that field.

This arrangement worked very well through its five year life, which saw the rise of Jeremy's business to it's peak, and then the beginning of the decline which ended in it's closure in the eighties. Prices in that trade were tight but at the beginning it was counter seasonal, about sixty percent of the business being, to Gardiner, spring season business, which helped greatly with that continuing problem.

At the outset of the arrangement came an illustration of the type of difficulty which we were having in our labour relations. After a long spell of Labour government only briefly interrupted between 1970 and 1974 by the disastrous Heath administration the shop stewards of our unionised workforce were feeling their oats and any change in working practices was, at that time, a matter of bitter argument. It mattered not that some changes only altered working practices, and sometimes even reduced workloads, but they all had to be bought. When, after our record long spell of short time working, the Ballantyne operation was proposed, we consulted the workforce and got their agreement to train on and operate four Somets in place of four Sulzers, which we sold, without any change in pay, the people recognising that the project would reduce the damage they suffered when on short time.

But it was not with surprise that we learned, when the machinery hit the floor, that the tuners didn't see it that way and it took many a long and bruising meeting before everyone agreed to co-operate. It was not until the Thatcher era that realism returned to industrial relations, and throughout the seventies Colin and I used to congratulate ourselves that our move into Sulzer weaving had been made before union power had developed at Gardiner, for the changes we made then would have been impossible in the late seventies.

It was at this time that I, who dealt with the labour problems, decided that early retirement, preferably at 55 in 1981, would be a good idea because for a matter of seven or eight years in that era I felt that management were not only fighting the market, the commercial enemies and competitors, but having to fight our own workforce as well. The licence generated in the Wilson years was a licence to use every possible pretext to get money, and if that seemed unattainable, simply to make life difficult for management. Whether the shop stewards should carry the blame for these attitudes, or whether the workforce as a whole had gone militant I'm not sure. Probably there was a bit of both, as the following examples will show.

The working arrangements in the spinning mill, agreed before it started up, included changing shifts on the run, with no stoppage of machinery. Stoppage was not just wasteful of productive capacity but was unnecessary and actually increased the workload, for starting up always produces a crop of thread or sliver breakages. In the beginning all went well, but one day, walking through the spinning mill at the time of the two o'clock shift change, I noticed that all the machinery was standing. On enquiry I found that I was the first to discover that this Spanish custom had developed, and it took weeks of argument and juggling of meal breaks to restore the position.

Seasonality in weaving meant, that even if the wages paid by Gardiner were accepted as reasonable a few weeks of short time could turn an acceptable annual income into a sub standard one and, having failed to eliminate the underlying problem, we sought ways of alleviating it. We proposed in 1980 to the stewards, and then to the workforce, a differential working week. This involved working 48 hours a week in five days in the first half of the year and 32 hours in four days in the second. The after tax benefit to a married employee with one child was calculated to be £451, a substantial sum when average wages were about £75 per week, gross. Shift working made the detailed arrangements complicated, but practical solutions were worked out for all the problems. We put the comprehensive scheme to the workforce, seeing a 70% favourable vote as the basis for implementation and did not get it.

An example of labour problems which would have been funny had it not been serious occurred in 1978. Our agreements with the workforce included a range of acceptable temperatures in the mills. In the bitter winter of that year the spinning mill Sunday night shift, which started the week, worked for the half hour which management was allowed to get the temperature up to the minimum acceptable, then stopped work claiming that the minimum had not been achieved. A thermometer was produced to prove it. It was not until the next day that we noticed that the little glass tag which holds the tube in place on the wooden scale had broken off, and the glass could be moved along the scale at will! It was very cold, and when the weaving shed started eight hours later we couldn't get it up to working temperature all day.

On a number of occasions, when an argument with the workforce was in progress, we were faced with the "phantom overtime ban". Despite almost universal shiftworking overtime was a regular feature of some operations, such as spinning mill maintenance. Several times we found that we could not get men to turn out for overtime. We knew that there was an overtime ban in force in pursuit of a current argument, but when questioned the shop stewards denied that any ban was in force. "It must just be inconvenient for the men in question. Overtime is not compulsory, you know".

It all made running a business in a declining industry that much more difficult. It was very stressful. When, in May 1986, both Colin and I retired, that particular pressure had gone away and I would have gone on more happily than I would have had retirement come in the late seventies.

George Lindores was due to retire in 1978 and we did not believe that there was anyone on the premises who could succeed him. More than that, we thought that it was time for a change of style. George was a Border bred and reared mill manager doing a sterling job getting out the product, but the flood of industrial legislation pouring out of Parliament did not exactly grab his attention and it seemed to us that his successor should be a different type of man and do the job in a different way. We started looking, and found that the decline of the industry had stripped it of most of it's likely young men and it was not until the tail end of 1977 that we found our man.

Brian Roberts was the son of Tim Roberts, a Director of George Roberts in Selkirk until it closed. Brian was trained at Leeds University, graduating just as the closure deprived him of his future. He went to a Kilmarnock spinning mill as quality controller, then to L. J. Booth & Co. Ltd., a firm of piece dye fabric makers in Horsforth near Leeds, on the sales side. After a few years there Booths fell on hard times and merged with another suffering firm, Newsomes. Brian was asked to transfer, but declined and went to Bradford University for a year to achieve an MBA Degree. Just as he was completing this we learned that he was about to come on the market, we interviewed him and on 4 October 1977 he started work in Selkirk as Mill Manager designate. When George retired the following year he succeeded him.

Ghosts from the past still haunted us. In the course of our problems with Mary Rae we, Colin and I, had each subscribed for 500 shares in her business in order to let us exercise some sort of control, and by 1977 the need for that had, with the decline of the Rae business with Gardiner, gone. We entered into an agreement for Gardiner to relieve us of our holdings with a view to them being sold back to the Rae family as and when possible. Deloitte advised that the transaction was either illegal or improper, and it did not take place and some years later Mary's son, Ewen, bought them from us when he joined the Rae business. Thus ended a long and exciting relationship!

The advantages of having two Companies, one buying and one selling, having long since gone, we decided at the beginning of 1978 to wind up one and operate under the name by which we were known to our customers, Gardiner of Selkirk Ltd. As Edward Gardiner & Sons Ltd. owned all the assets we had to have the Companies change names before the old Gardiner of Selkirk could be wound up, which we achieved by first changing that Company's name to Lesgar Ltd. to clear Gardiner of Selkirk off the register. Then we changed Edward

Gardiner's name to Gardiner of Selkirk Ltd., then we wound up Lesgar Ltd. All this took, with statutory notices, meetings, advertisement in the gazette and the like, most of 1978, but by the end of that year Gardiner of Selkirk Ltd., stood alone. I had been discharged as liquidator of Lesgar and everyone was happy. Except the Inspector of Taxes in Galashiels who, two years later, summoned the non existent Lesgar to appear before the commissioners for failing to submit accounts!

The West German market, in which Gardiner was prospering on the back of Bill Rohe's efforts was one in which garment makers were using very up to date machinery and demanding higher and higher stability in the fabrics which they bought. This meant that more and more of the product had to go to Galashiels to have the Gladstone "London Shrunk" process applied. Similarly, our biggest tartan customer, Moffat Woollens, also called for this finish. Though not the whole of the London Shrinking process, heat setting in a decatising machine was the part of it which imparted fabric stability, and the machine which did it was available from Sellers, of Huddersfield, and in November 1977 the Gardiner Board decided to buy one for £49,000, financed by a term loan from the Royal Bank at 4% over base rate repayable over two years. The supporting papers showed that even at two thirds of current volumes the payback time would only be two and a half years, assuming that we could build into our prices the figures charged to us by Gladstone. The venture had the added advantage of shortening delivery times, for sending cloth out to a contract processor inevitably costs valuable time.

One of the underlying reasons for the spinning project had been the desire to have better quality yarn, for high speed weaving on Sulzer machines is more demanding of yarn than is conventional weaving at less than half the speed and, as the trend to lighter and lighter fabrics continued it's relentless course, both winter and spring, benefits began to flow through. Blends had been changed, dramatically in the case of the finer S24 and S30 qualities. The improved quality and strength of the latter made tartan production much more efficient, and the better S24 yarn allowed the manufacture of fabrics, single yarn plain weave cloths being a typical case, which we had hitherto had to refuse to make because of their poor weaving performance. It is very probable that had the spinning mill not been built the accommodation to lighter weights demanded by the market by the end of the nineteen seventies would not have been possible.

Computer development in this era was fast and furious. The ICL 1901 which we used was a card input batch machine and, by the middle of the decade, keyboard input machines with access to more than one programme at a time were available. One of the leading machines was the American Singer System 10 whose manufacturer became an ICL subsidiary with the machine becoming the ICL System 10. In late 1978 we decided to convert from the 1901 to the System 10 at the summer holiday of 1979. We ordered the machine, and commissioned, again through ICL, it's programming, taking the fullest possible advantage of it's new capabilities.

The bookkeeping programmes were bought as standard packages, as was the wages programme, but stock and production control were to be our own, and in this field we made the biggest change of all. Hitherto yarn requirements for cloth orders had been calculated by hand, punched into card and fed into the machine to update the stock position. Similarly issues of yarn against the requirements thus created had been entered item by item. The enormous change now possible was to have all the particulars of each colour of each design in store, awaiting receipt of orders. When they arrived, were they for sample lengths or bulk requirements, the design, colour and quantity were entered and the machine, using a series of linked programmes, not only calculated the yarn requirements and updated the total requirements by required week, but went on to print all the weaving and finishing production documents for the order, suitably broken down, where necessary, into logical weaving beams. When the time came to issue the yarn to production simple identification of the beam concerned triggered the issue, and the removal from outstanding requirements. It was a tremendous step forward, but once again our testing was inadequate and the summer holiday start up failed. The bugs grew faster than we were killing them, we went back to the 1901 and tried again, this time successfully, in November. The properly tested programme is akin to the pot of gold at the end of the rainbow.

Whilst in the spinning mill's early days we had been continuing to take yarn from Pogson and Shires the trend to lighter weights and the increasing difficulties of the weaving trade had, by the end of the decade, eliminated the need for such purchases. Indeed, as early as 1977 the spring season gap was such that spinning or weaving production and stockbuilding could not occupy the mill from September through to January and work had to be found for it. David Stevenson, at the Edinburgh Woollen Mill in Langholm, provided the answer which gave us a welcome breathing space. He was expanding rapidly and in the time of the tax concept of Stock Relief it suited him to hold increasing stocks over his year end in the autumn, so long as they could remain on our premises. The yarn of which he used most at that time was 2/8s Shetland, ideally suited to our machinery, so we entered into substantial contracts with EWM which went a long way to plugging the spring gap in the spinning mill, and which by the negotiation of further arrangements to weave a large part of the yarn so produced was a great help to the weaving problem in the second half.

An arrangement as perfect as this could only have a limited life and was upset by two unrelated factors. When stock relief ended the incentive to overstock at the EWM year end disappeared and, inevitably, crystal balls being cloudy at the best of times, yarn was required at our busy times to plug the wrong forecasts of the Autumn.

With the success of the EWM spinning operation behind us we began to sell yarn to other customers. These firms did not offer the off season benefits provided by EWM and their demands tended to go with our own seasons rather than counter to them, so we settled down to the next great debate. How to cope with a year round sale yarn activity.

Chapter Twenty
Spinning Dominant 1979/1980

A cursory glance at the table of results in the previous chapter reveals the fact that, from the day the spinning division became profitable, it's results overshadowed those of the weaving division. The contrast is muddied a little by certain accounting policies, and inevitably hangs on the transfer price of yarn from one division to the other.

In the early days, when all the yarn produced was consumed by Gardiner's own looms, the transfer price was critical, for it was the price used in costing woven fabrics; an excessive transfer price would both make the weaving division's struggle for profitability that much more difficult and inflate the spinning result. Conversely, too low a price would both hold back spinning profitability and artificially depress cloth prices if compensation in the form of increased cloth margins was not applied, and increased margins in cloth operations were something which all our experience inhibited us from seeking.

In the later days, when yarn was being sold to outsiders, the concept of the spinning mill as producer, going for a proper return on capital on it's transfer price, was continued, with all costs and benefits after transfer accruing to the weaving division. Thus selling prices to outsiders had to have built into them a number of items. Much yarn was sold in twofold form, so the twisting cost and the cost of winding onto the ultimate package had to be added. Selling agents commission and selling and administration expenses, along with travel and other expenses had to be built in, along with a margin of profit on all these added burdens. As the sale yarn business developed budgets had to be developed to ensure that, when sales of yarn became substantial, we did not find ourselves with costs unrecovered from our customers.

Thus, as the sale yarn business developed, the weaving division result included an element of profit which was related to yarn processing and merchanting which, over time, became substantial. The monthly management accounts derived from a comparison of each division's expenses with a figure which we called "manufacturing contribution", which, in effect, was the sales value of production with all raw material costs, agent's commissions and standard cash discounts stripped out. This system of accounting took no account of fluctuations in levels of stock and work in progress and reflected none of the variations to which a system incorporating actual sales figures and stock levels is prone. The method allowed the identification of contribution applicable to sale yarn activities in the weaving division and, typically, for December 1987 the sale yarn contribution was £19,000 against a straight weaving contribution of £103,000, a significant 16% of the division's total. When that very successful year ended a month later the percentage was 14%.

Even so, the incorporation of sale yarn activities into the weaving figures should not be seen as too distortive, for, at the extreme, had there been no contribution there would have been an almost balancing reduction in expenses, for only the profit on post spinning processing accrued to the weaving side.

So, if the allocation of profit between the sectors is accepted as rational and reasonable the message conveyed by the figures is clear. And confusing. Of the two operations spinning was the more profitable, with weaving struggling to make a profit at all, and often losing in the second half a very large part of what had been made in the first. Any modern right thinking manager would close the weaving and operate just a sale yarn spinning business, but only if he knew that the very substantial, much more than half, part of the spinning output which was being used in cloth production could be sold outside. The industry supplying the knitwear trade with yarn was itself a very competitive one which could not be broken into massively and easily.

Management's theoretical view of weaving had, therefore, to be tempered by this background and continued to be so tempered for the remaining years of this history. Even at it's conclusion, after the year to January 1989 when 56% of yarn produced was sold on as yarn, weaving remained a core element of the business, producing 47% of the profit.

In the period 1975 to 1980 records of management's deliberations are thin on the ground. With the withdrawal from the Board of the Tootal representatives the need for formal meetings declined and from the frequency of about nine a year we slipped for a while to a rate of three or four a year and at one point went from September 1975 to March 1976 without a formal meeting. Nor is there a record in correspondence between me and Colin and Tootals, for their divorce from us was absolute, the only contact being occasional meetings with Norman Hornsby. There was no point is sending monthly figures to Manchester and the only information provided to Tootal was the annual accounts.

The absence of formal meetings should not be taken to imply an absence of discussion and debate for in a rapidly changing industry debate and survival are inseparable. The fate, or continuation or closure of the weaving division was the subject of continuous argument both for the commercial reasons mentioned earlier and for it's sentimental background both as the starting point of the Gardiner business and as the employer of a lot of people that Colin and I had known and worked with for a very long time. Perhaps there is a commercial argument for applying some sentiment to business, for had we simply looked at the figures the weaving division would have gone and would not have been available to make its massive contribution to the results in the latter half of the eighties.

From the opening of the spinning mill in 1976 to the end of the decade weaving activity was dominated by the Ballantyne furnishing business, Bill Rohe's activities in West Germany, the very substantial Edinburgh Woollen Mill contracts and our London Men's wear agent Alf Masterman's development of the Burberry account. Between them they filled a good three quarters of our capacity and balancing their conflicting demands required both the capacity statements churned out by the computer and their reconciliation with the "capacity reservations" made by them and, indeed, by all our other agents and major customers. Each Friday a statement was made up setting out the position on the basis of which delivery offers for future orders were adjusted. Space reservations were made without payment, but the customers making them regarded them as firm promises on our part, but discardable options on theirs, an extremely uncommercial concept from the manufacturer's point of view. Naturally any suggestion that options should be bought not given simply produced derisory laughter. It was in all respects a no win situation as we found to our cost when a major customer who habitually placed most of his orders in the autumn found demand running away from him and wanted cloth far in excess of his reservations in our peak period. Gardiner was deemed to be at fault and the problem ultimately led to the splitting of a lucrative exclusive contract with another manufacturer.

The year 1978 saw a number of changes at departmental manager level. George Lindores, whose successor Brian Roberts was in place, retired after a long and very productive career. Earlier in that year George Watson, manager of the pre weaving departments and one of the industry's great characters, retired because of ill health and did not live to enjoy his retirement. He it was who used to tour the mill when the Christmas parties were in full swing and deliver a semi humorous, semi serious, speech concerning the year just past and the prospects for the future. I think the workforce set more store by George's predictions than by the official ones which came out through the works committee reports! His replacement, Tom Spence, came to us from Ballantynes of Walkerburn, partly driven away from them by disillusionment with the performance of their computer programmes.

That same year Gordon Hardie, the weaving manager, was replaced by Ian Cooper. Gordon was probably the best tuner in the Borders, but his heart was in country pursuits so when we decided to replace him by a different style of manager he went off to live the country life with no heartbreak. Ian Cooper, a young man who had served his time at George Robert's Forest Mill and who had joined us as a tuner when our Sulzers were introduced was the new weaving manager. From a different mould than most tuners he is a thinking man who appeared in my notes on one of our disputes with the tuning force as follows; "I think someone is stirring it, I think it is Ian Cooper". Ian left us to go to South Africa to manage a large weaving shed in that country. For some reason he and I corresponded in a desultory way and when he decided to leave South Africa he agreed to come back to Gardiner as weaving manager.

John Knox had been finishing manager since returning from war service in the Air Force, and retired in 1987. His immediate successor did not match our expectations and in 1979 was replaced by Rob Howden, still in place at the end of this history. Within the mill this replacement of all the senior managers in so short a time could have caused difficulties, but the quality of the new men allowed the changes of personnel and style to go in smoothly.

At the sharp end, in London, Ken Smith retired from Alan Shepherd Ltd. in 1978. Whilst his expertise and deep knowledge of the trade was to be missed his retirement was in some ways fortunate. The London market was in severe decline, with the substantial customers of the past disappearing the scene or changing their product away from our fabrics, and the new firms not replacing them fast enough. By this time the Marks and Spencer skirt business had virtually disappeared and our sole continuing fabric was a raised pure new wool cloth for childrens coats. It was eminently suitable for manufacture by the Yorkshire specialists so we had to fight hard to keep our share of this business. All these factors meant that a reduction in scale of the Shepherd operation was essential and Ken's retirement provided the means. He went to live in Bognor Regis and played a lot of golf while Gerald Box got rid of all his staff and sub let part of the office. From being the driving force London women's wear had become a fringe market again. From being 95% home trade, Gardiner's cloth business had become close to 50% export, with Germany, Canada and the United States in the driving seat.

Our engagement of Alf Masterman to conduct our flagging men's wear business had brought in a number of valuable accounts the most important being Burberry, who used our fabrics for sports jackets, skirts and, most importantly, for "warmers", the button-in lining for their famous raincoats. First under their buyers Bob Sundler and later, after his retirement, under his successor David Hall the business grew and grew. It came, therefore, as an unprecedented shock when Alf, whose business with his other principals as well as with us was flourishing, committed suicide in 1979. We didn't replace him, conducting the business directly from the mill, but he was sorely missed.

On an entirely frivolous impulse the Board decided that Gardiner should invest in the 1979 challenge for the America's Cup, the

Lionheart challenge. The syndicate involved appealed for 1500 industrial firms to put up £1,500 each to fund the venture and the Gardiner Board, feeling modestly prosperous at the time, signed up. For many firms there were commercial spin offs in the form of publicity for products involved in the challenge, but our involvement was "just for fun". Only one other Scottish firm, a knitwear manufacturer in Dumfries, signed up and they provided some of the knitwear for the crew. There was no collaboration but it is an odd coincidence that Hamish Robertson of that firm and I both sailed with the Solway Yacht Club!

By the year end in January 1979 the balance on the Profit and Loss account had reached £726,000, most of which was, realistically, undistributable, so the possibility of increasing the Ordinary Capital to £800,000 by a bonus issue of half of that amount was mooted. In July it was rejected, the Board feeling that the fees involved would bring no commensurate benefit.

Meanwhile, despite the lack of formal meetings, the arguments about the relative importance of the two divisions rumbled on. The idea that the weaving might be completely closed had not yet taken root, but it's size and operations were under constant review, not least because of the increasing difficulty of the spring, second half, season. That worsening problem was reflecting in the spinning division's second half activity, making it more and more difficult to keep the plant going in the autumn. Whilst the winter, first half, season called for almost all the yarn produced to be used in weaving, the second half was a different story, and sales to outsiders were a necessity. Barring the special, temporary case of the Edinburgh Woollen Mill, knitwear customers wanted year round capacity, with a seasonal bias to the first half, a demand which our two set mill could not accommodate.

Launching into a full scale sale yarn activity, a venture which would require expansion of the spinning mill, in the face of competition from the established producers of Shetland yarn, Laidlaw and Fairgrieve, Knoll Spinning, RTN, Yarns and the like, was not to be undertaken lightly, so we sought advice from the International Wool Secretariat (IWS) on the probable future market for that product. One of their economists, Juliet Bestoe, came to see us, learned what we wanted to know and a few weeks later came back to report. The IWS advice was that the market was big enough to accommodate us, so on 12 November 1979 the Board decided to buy an 85" Tatham Carding Machine, two HDB spinning frames and appropriate ancillary equipment, a total spend of £400,000. Delivery was of the essence, for Industry Act grants were being phased out, and plant not operational by the end of June 1980 would not qualify.

Because we had built the building in 1975 to the original, three set, planned size the new machinery could be fitted into it, the only major construction work needed being the digging of the pit over which the new machine would stand. A temporary wall was built and the excavation and construction of the pit was achieved with no interruption to production. Whilst grants would provide £150,000 of the capital needed the working capital which would be needed for the new business meant that somebody had to be found to finance the plant. HDB took on the financing, over five years, of the spinning frames and though Tathams declined to fund the Carding Set they introduced us to Hambros, who did so.

In the run up to the November decision we had thought it proper to advise Tootal of what was in our minds and they responded with a visit by Alan Wagstaff, by then Chairman of the Company who, having looked over the figures and projections provided to him, bluntly suggested that the proper way to fund the operation was by closing down the weaving activity altogether. This was the first time that the idea had come into the light of day. Colin and I agreed to consider it while expressing our reluctance to take such a step, and one of the results of that consideration was an addition to the papers which the Board discussed in November in the form of an additional cash flow projection based on weaving closure.

The working capital requirement of the sale yarn activity was put at £283,000 if it ran in harness with weaving, and £846,000 if it were the sole activity. Naturally the abandonment of weaving would eliminate that activity's demand for capital and the sale of the plant would bring in further funds. The Royal Bank had agreed to an overdraft limit of £669,000 which more than covered our needs with weaving continued so we did not feel under pressure to go along with the Tootal proposal notwithstanding the dramatic difference in the projected cash flows. Condition one, which was weaving continued, showed an overdraft peak in year one of £463,750 going into credit in year four to the tune of £180,000. But those were the days of peak inflation, and the final, inflation adjusted, line of the projection showed a year four overdraft of £671,000 and even by year seven the overdraft would still be a hefty £338,000. On the other hand condition two, the projection for weaving closed, forecast a million pounds in the bank at the end of year four. No wonder Tootal favoured the end of weaving. But in the end we decided to keep our basic business going, a decision which we sometimes regretted in the years between 1979 and 1986.

With a subject as explosive as weaving closure under discussion keeping things secret was quite a problem. The minutes of these meetings were written and typed by me at home, and the "condition two" paper is hand written and photocopied and the only copy extant is that appended to the minute. For once we managed to achieve secrecy for, to the best of our knowledge, no hint of the condition two proposal ever got out.

The machinery was all ordered, was all delivered to time, was operational by the end of June 1980 and we were able to collect all our grants. But, as was the case when the spinning mill was first opened, trade was poor, and once started up the new set was shut down and not brought into production for nearly two years.

By the time that stage had been reached we had appointed two Leicester men, Geoff Tanner and Ian Elms to sell our yarns in the provinces. We intended to handle Scotland ourselves, which we did for a number of years until we appointed Ronnie Allan to the job in 1983.

The year to 31 January 1979, the last before the decision to move into the sale yarn field was taken, was profitable for both divisions. The first half of 1979/80 was much less profitable than the year before. All the doubts about weaving were well founded, for in the year to January 1980 the division fell into loss for only the second time since 1962. And the intervening loss had not been a real one, so it was really the first time in eighteen years.

Chapter Twenty One
To Weave or Not To Weave 1980/1982

Between 12 November 1979 and 10 October 1980 no Board meeting was held, matters of policy being dealt with over the morning mail and at the meeting at which matters arising from the mail were settled. The morning meeting, held immediately after the mail had been opened and involving the Directors, the Mill Manager and the heads of departments, had been a significant event for more than twenty years.

In it's early days a large part of the daily post was the letter from Alan Shepherd in London, and the order forms which came with it, so we had a shorthand typist sitting in to allow us to answer the letter on the spot as the decisions about it's contents were taken. As time went by we came upon a period when the current typist suffered frequent absences for sickness so, the early tape based dictating machines being available, we equipped ourselves with them to eliminate the need for shorthand. The stenographer never had a day's illness after that but needless to say she had the same workload but delivered to her on tape instead of by direct word. As time moved on and telex was installed the mail to be dealt with declined dramatically, the majority of orders arriving by telex and being dealt with through the day, but we continued the meeting as a useful daily gathering at which problems and ideas beyond the immediate mail could be aired.

Robert Rae lived to see the third set commissioned but died in September 1980, having played an invaluable part in getting us into the spinning business. At the Annual General Meeting on 10 October 1980 Brian Roberts was appointed to the Board.

To the increasing problems of Winter/Spring imbalance there were now added those created by the severe recession brought on by the Thatcher Government's determined efforts to reduce inflation. Hindsight insists that the weaving operation should have been severely cut back after the summer holiday in 1980, for the weaving division was to lose £115,423 between then and the year end, but we didn't do it and paid the price. But by the Autumn it was plain that we could not carry on a weaving business, year round, on the scale at which we had operated hitherto and on 3 November 1980 the Board, looking ahead to the end of the next winter season, approved 7 January redundancies and a further 70, including the weaving night shift, for end August 1981. This was to be the first of a series of adjustments to the weaving division to be made over the next few years. The same minute notes that the manpower plan for the extended spinning mill had been put to the Union but had not yet been put to the workers. It was to be the subject of a long drawn out wrangle.

The original plan for the spinning mill had called for wool bins at the hopper ends of the carding machines to house the blends in production, but this had been abandoned in the design stage and the practice of baling all blends, and working from bales, adopted. It was by now apparent that bins would be a substantial advantage and four bins each of three tonnes capacity were ordered to sit behind the three carding sets, together with the fans and pipework to fill them from the blending plant. Bin capacity follows Parkinson's Law. No matter how much capacity is provided it is never enough, so we frequently had to resort to baling when things got out of step.

By November 1980 the winter season of 1981 was under way, with the first bulk Marks and Spencer order in hand, and weaving returned to five day working. Spinning, having spent the autumn building yarn stocks and currently benefiting from the State short time subsidy, stayed on short time until the subsidy ran out.

The Eyemouth factory was for sale but not selling. An offer from a fish processing firm fell through, leaving us with the problem of whether or not to give it the coat of paint which it needed.

Though the November decision had set the pattern for a cutback in weaving the underlying argument, to weave or not, went on like a nagging ulcer and the December Board meeting, having noted that provided full activity could be maintained through December a profitable year was possible (in the event a profit of £66,375 was earned), the minute went on thus:−

" It was agreed that, whilst with the sale yarn spinning trade in it's present state of development the abandonment of weaving at the end of the season Winter 1981 could not be contemplated, a reduction to a Sulzer and narrow Somet configuration early in the New Year, and to an all Sulzer configuration at the end of the season, was desirable".

The Export Group had got all excited about the prospects of selling yarn in Hong Kong and I had taken advantage of their package to go there to explore the prospects. I found them to be minimal for our product, but went on to New Zealand to learn about buying wool direct from brokers in container loads, rather than from wool merchants by the bale. This seemed to be advantageous provided that substantial quantities were still bought from merchants, for boxes meant that, in the early days, 7° tonnes landed in the woolstore all at once and, later, as bale pressing machinery grew more powerful, fifteen tonnes arrived all at once.

As 1980 gave way to 1981 and the closing down of the wide Somets, no longer needed as the Ballantyne contract worked itself out, became due, it was clear that the changes in work practices which would ensue would, with the workforce in it's present mood, cause a lot of problems, so we kept them going. In the spinning mill the bonus scheme which we had put forward as part of the manning plan for the third set had struck rocks, and a revised plan had to be thought out. The freedom of action which Trade Unions had enjoyed in the Wilson years had not yet crumbled away; even the current recession and trade difficulties had, as yet, made no impression on attitudes. And still the debate on the size and format of the weaving operation went on, with the January meeting discussing not only weaving but it's pattern making, pattern cutting and administrative support, for it made no sense to cut back the productive plant leaving the support services intact with their costs undiminished.

March 1981 saw the approval of the years accounts and set a 2p per share dividend for payment on 30 April. Though by now we were moving in and out of credit at the Bank, trading conditions did not give us confidence enough to rattle up the dividend. On the spinning front an agent, a Mr. Pattinson, was appointed to sell our weaving yarns in Yorkshire, where we hoped there might be a market in the weaving trade, and we decided to show at the Leicester Yarn Show in the Autumn, not taking space in the Spring Show. The Eyemouth factory was put into the hands of Bernard Thorpe and Partners for sale, for we were having no success in selling it privately. At that time, March 1981, everything was more than a little muddled and cloudy, and, as a Board, we hardly knew where to turn. All we could do was press on with the development of the spinning operation and watch the weaving division like a flight of hawks. To further the sale yarn operation we decided to buy a semi automatic machine for strapping yarn cartons, and set up a packing line in a now disused part of the old yarn store.

This was the time of the fashion for Current Cost Accounting, a cause enthusiastically espoused by Jack Shaw, Senior Partner in our Auditors, Deloittes. They held seminars and teach ins in an effort to convince their clients but, for all the weaknesses of conventional accounts in inflationary times I concluded, firstly, that there was an element of promotion of the accountancy profession in this enthusiasm, for if all companies adopted CCA the profession did not have the manpower to do the job and, secondly, a balance sheet and profit and loss account in which the only real, undoctored, figure was cash did not seem to me to be a good idea. And I had no doubt that the profession was working on a way to doctor the cash.

So, to keep the Auditors at bay we passed a resolution to the effect that we considered CCA to be fundamentally inflationary, time consuming, expensive and irrelevant to the Company's needs, and we were not going to play.

Development of the sale yarn spinning project continued. In May 1981 Helga May was appointed agent for yarn in West Germany and our first contacts with Dr. Tessari, who became our very successful Italian agent, were made. Applicants for the agency in the United States were short listed to four including Jack Donovan who was eventually appointed, and unsuccessful. And Colin and I spent many a day on the road supporting our men in Leicester, Geoff Tanner and Ian Elmes, in their efforts to sell to the big names in the knitwear industry. Corah, Nottingham Manufacturing, Albert Martin (Cooper and Roe) and the like. In the event we were to have as much or more success with firms a little smaller such as Dukes and Marcus and

Glover, but minimal results from the even smaller firms lower down the scale even though we offered, in our main Shetland quality, a stock service in some 48 shades. In the case of Dukes and Marcus there was an added benefit in that one of their principal customers, British Home Stores, sought skirts to match the knitwear and a useful cloth business grew out of that. The wide Somet problem was solved by selling them and we overcame the manning problems which resulted.

With the end of the short time subsidies and the effects of the redundancy payments act and new legislation forcing payment for the first several days of short time in any one three month period the balance of advantage between going on short time and paying people off became not a little confused. Confusion or not none of this legislation, which placed barriers across the traditional route to cost saving in a seasonal downturn, short time working, made life any easier. It was galling to pay a man off, make a redundancy payment, and then a few months later to take him on again as trade turned up. And not only to management, but to the Unions too, for they insisted that we agree that a worker choosing redundancy could not be re-employed. This was an impracticable condition, and fell out of use quite soon. Too much labour legislation can have unforeseen effects, and the combination of rules noted above led to the situation that it could easily be cheaper to pay off the newest entrants at the onset of a downturn rather than cope by short time working, an effect unforeseen, I'm sure, by the brains of Westminster.

By May 1981 the problem of shading, or colour matching, the host of new colours which we had to make for our new knitwear customers had become serious, so we shifted one of our darners, who had shown an aptitude for the work when on loan "helping out" to assist Colin Ford on a permanent basis. This created an accommodation problem which we solved by buying a Portakabin and mounting it on brick pillars in the woolstore. Thus elevated it was connected to the existing first floor shading department, and used no woolstore floor space. And with the problems of sampling to knitwear customers without disrupting bulk production needing urgent solution we began the search for a sixty inch carding set upon which to card small sample batches. For all that times were still very difficult, with weaving on three days a week in the early winter with the Board noting that it should have been two days, and spinning working a sustainable three days a week, the year as a whole and particularly that of the spinning division, was shaping up quite well. Weaving and spinning had made £123,000 and £105,000 respectively in the first half, and while the former was to lose £31,000 of it's first half gains in the second, spinning was to add six and a half thousand to it's tally. The year's result of £235,000 was to be, in the circumstances, eminently satisfactory.

This worthwhile profit when added to the hefty £167,000 depreciation charge allowed Gardiner to repay all machinery loans but that on the last carding set and the building and, leaving cash in the bank, to buy £168,000 of Treasury Stock yielding, in those high interest days, 14⅝%, which was sold five months later for a profit of £15,000 when the working capital demands of the winter 1982 season put the Company back into overdraft.

For the year to January 1982 the Board felt able to increase the dividend by 150%, to 5p per share.

When the spinning mill was being planned a deliberate decision was taken, prompted by the twin goads of finance and the launch into a technology entirely new to the Company, not to consider building a dye works to dye the wool we would consume. That function was left to specialist dyers, principally Brook Dyeing in Holmfirth and, to keep part of the operation close to home, to Jack Cruickshank's Bridgehaugh Dyeworks in Selkirk, dyeing about half each.

We knew from the start that Bridgehaugh were more expensive than Brook, but accepted it so as to keep some of the work in Selkirk, but by 1981 the price gap had become so large as to cause us concern. We thought about building a new dyehouse behind the spinning mill, and discarded the idea, and then began wondering whether, if more dyeing was put to Bridgehaugh their cost structure might change to reduce or eliminate the cost gap.

After trying to estimate the effects and reaching no conclusion we approached Jack with the idea that if we jointly explored the proposition, and it worked out favourably, we might, if he agreed, buy or take a majority interest in his business, and Jack was only too willing to co-operate for the reason for the enquiry was as obvious to him as it was to us.

For a month we delved into Bridgehaugh's costs, capacity, potential capacity and everything that might happen if we pushed more work their way but at the end of the day we could not see their costs getting down to the Brook level. So, reluctantly, for all parties knew that it meant a reduction in the Gardiner/Bridgehaugh throughput, negotiations were broken off. We continued to send work there, but not as much as before. A year or two later, Bridgehaugh was bought by the Edinburgh Woollen Mill. This surprised us.

Efforts to introduce more flexible manning arrangements in the weaving division continued to be made without any progress being achieved. Proposals put in November 1981 were rejected, and more were generated for discussion in February.

In November the Eyemouth factory was sold for £15,000 for use as a motor showroom. In January 1982 Tathams found a 60" carding set taking off 120 ends at Harvey Rhodes, in Delph, which was closing, and offered it to us, reconditioned, for £29,000. We accepted, and arranged to erect it in the wool store, keeping it separate from main stream production, for sampling only.

Through 1982, despite continuing, successful, efforts to develop the sale yarn spinning business the weaving shed dominated management thinking. In April our agents were persuaded that in the absence of much proper spinning business they should attempt to obtain bulk, fill in, business at prices substantially lower than list, for even with the Edinburgh Woollen Mill business there was still a yawning hole in that season. The logic seemed sound, for the EWM business, though concentrated in the off season, was offered at prices set by the stars of the Yorkshire commission weaving industry, and very tight prices they were. The Board reckoned that the 20 Sulzers, of 22 owned, that were currently working double day shifts were an appropriate complement for the busy winter season, but notwithstanding the EWM business were too much for spring and short time working was therefore inevitable in that season.

That conclusion was recorded in the minutes of the April Board meeting in ignorance of the blow which was about to fall upon us.

Shortly afterwards rumours began to circulate in Selkirk that Edinburgh Woollen Mill had bought Heather Mills, the second biggest, next to Gardiner, weaving mill in Selkirk; one which also had a spinning plant. That was important to us, not only because if it were true we would surely lose all EWM's commission weaving, but we might lose a lot of spinning as well, for we were, in the main, weaving yarn which we had made and sold to EWM.

On enquiry David Stevenson confirmed that moves were afoot, but that the deal hung on Heather Mill's performance in the months ahead. If it performed to certain standards, it would be bought. If not, the deal would not happen. Our present year's weaving was secure, but after that, if the deal went through, things would be different. A very fraught few months, over which we had no control at all, lay ahead of us. Even with that weaving the spring season lost money. Without it it would be a desert.

In due course the deal did go through and three thousand pieces a year of off season weaving left Gardiner for a new home down the road at Heather Mills. But every cloud has a silver lining, and we managed to hold on to the spinning contracts, our modern plant being designed to produce the range of counts into which eights Shetland fitted very efficiently.

In September 1982 we were still hoping to replace the lost weaving by cut price offers abroad, but by late November we had become reconciled to only limited success in that effort. The 22 November meeting took an even gloomier view of prospects and recorded a rather drastic conclusion; "The weaving division prospects for the first half of 1983 were reviewed. It was agreed that provided reasonable orders from Marks and Spencer were forthcoming a two shift twenty loom configuration to be operated to July. However, the season in prospect looked by no means brilliant, and it was agreed that, in the absence of effective amelioration of the extreme seasonality of the Company's trading pattern a decision must be taken no later than March 1983 whether or not to close down the weaving division at the end of July 1983".

In December we learned that our Marks and Spencer contracts had been severely cut and the Board decided to reduce the weaving to single shift, with drastic changes in manning and work practices. These new arrangements, to be described in the next chapter, were to form the platform from which the weaving division was to fight it's way back to prosperity, without EWM and with or without Marks and Spencer.

The latter, which had been vital to the bringing of the weaving mill to true prosperity, no longer used our woollen fabrics for skirts, and we hung onto only part of the childrens coating contract, the bigger part of it being woven in Yorkshire. We had, in fact, lost it a year or two earlier, only to be reinstated when twenty thousand coats, made from a competitors cloth using shorter fibre than used by us, failed rub tests and were rejected.

Our relationship with Marks and Spencer had always been a good one; we might not have hung on to the childrens coating cloth for so long had it not been, but it did have its fraught moments. Fraught and amusing, from some points of view. One day Colin was negotiating the price for a contract, making a great virtue of our continuing last years price, and said, "if you can persuade the garment maker to go on at his last year's price, so can M & S maintain their retail price". "Not so" said the selector, "we are looking to increase our margins, so we want a better price from you".

We were not alone with our spring season problem. The whole industry was affected to a greater or lesser degree. So much so that the Chairman of the National Association of Scottish Woollen Manufacturers called a Conference in Edinburgh to discuss the problem. The Conference was well attended, and was chaired by Ian MacIntosh, of Reid and Taylor. After an afternoon spent getting nowhere Ian turned to Edwin Armitage, a Yorkshireman whose firm, the Woolly Mill, was in Langholm, and said "Edwin, we haven't heard from you yet. Have you any ideas?" To which Edwin, who never minced his words, replied "No, I don't know the cure either. And if I did I wouldn't tell you lot". Which brought the Conference to an ineffective and hilarious end.

Chapter Twenty Two
Time to Grasp the Nettle 1982/1984

Whatever the view of the desirability of continuing the weaving division it was now perfectly clear, without any fancy arithmetical analysis, that it could not be sustained at it's present size, or even in it's present shape.

One of Ewen McNairn's early moves had been to separate pattern making from main shed production by setting up an entirely independent pattern making department. This section made both the blanket ranges which made up the collection used for selling purposes, and also made the sample lengths demanded by customers before they placed their orders for full pieces. This was a logical and effective way of doing things, for as a season's collection was nearing completion the first of that season's sample lengths would be coming in, and as the pressure of sample lengths began to wane the making of the next season's ranges would be gathering pace. For many years the pattern shop had benefited from a great deal of overtime, a situation not appreciated by the main shed when they were, themselves, working short time, but sampling and rangemaking were the seed corn of the bulk orders.

The decline of the spring season had brought about a reduction in the size of the spring range, and a severe decline in the demand for sample lengths, so the pattern department was, by 1982, experiencing seasonal imbalance, imbalance which, whilst not precisely counter cyclical to that of the main shed, tended that way. Any review of the whole weaving picture had to take this into account.

The three Directors, Colin Brown, Ian Jackson and Brian Roberts, who met on 17 December 1982, knew that unpalatable decisions about weaving had to be made and could be delayed no longer, and they had before them a detailed analysis of the seasonal picture with

Edinburgh Woollen Mill and Marks and Spencer orders lost. It was not a pretty picture. From a peak of more than twenty thousand pieces a year not so very long ago Gardiner was looking at a probably yearly total of just a shade over six thousand, with a dramatic concentration of orders into the first half of the year.

The detailed analysis of week by week orders for the winter 1982 season, in terms of delivery weeks, set against a single shift capacity of 155 pieces a week without overtime, or 180 with, showed that there were only 11 weeks in the year in which demand exceeded single shift capacity. In those weeks the excess demand totalled 1200 pieces. From week 7 to week 29, the last week before the summer holiday where demand did not exceed capacity, the average demand was a mere 104 pieces, adding up to a shortfall against overtime capacity of 808 pieces. Add to that the shortfall against capacity of 384 pieces in weeks 1 to 6 and single shift capacity and demand were seen to be nearly in balance, FOR THE WINTER SEASON.

From the summer holiday onwards, the spring season, the picture was very different. Only in one week, when 158 pieces were required, did demand exceed ordinary time capacity, and the 19 week season totalled 1285 pieces, an average of only 68 per week. During that period, even with the remains of the EWM work, the main shed had been on short time, while the pattern shop worked full time. It was clear that much more than a mere reduction of capacity by half by adopting single shift working was needed.

Six thousand pieces generated a yarn requirement of 180,000 kg of yarn, or 14 weeks of spinning mill output, and the sale yarn business was not yet strong enough to sell the other 32 working weeks product to the knitting industry, so the Board concluded that weaving must continue at least until spinning was ready for "free flight". But on very different terms than had prevailed hitherto. The proposed solution was drastic, detailed and was not amenable to modification by the shop stewards on behalf of the workers. Whilst over the past 25 years we had assiduously tried to avoid "either or" propositions this was the one which had to be put in that category. Either this, or weaving stops. Enormous changes in working practices were built into the proposals, and all the flexibility which we had sought, and mainly failed to achieve over the years, had to be asked for, and delivered.

The plan for the weaving shed was as follows:-

Overall capacity to be cut back until it fell short of winter demand.

Supporting services and overheads to be cut back as severely as possible. Where doubt existed, cut.

Eight Hattersley looms used for pattern work, the Hergeth sample warping machine, the sample weft winding and sample twisting machines all to be moved into the main weaving shed. Space had been created in the reduction from 26 to 20 machines.

That total flexibility be required of the weaving shed staff, working any combination of the 20 Sulzers and 8 Hattersleys which might be appropriate from time to time. Flexibility to be extended beyond the weaving shed with finishing machine operators to run whatever machine is required at the time, winders moving between warp and weft winding, designers doing their own clerking (much reduced since the computerisation of production documents). Redundancies would be called for which would reduce the weaving division workforce, including clerical and computer services, to 80 people.

The minute of that meeting calls for the machinery movements and the implementation of the proposals as early in 1983 as is possible. The opening question in the Board paper had been answered. Weaving was NOT to close at the summer holiday 1983.

The underlying objective of these somewhat drastic proposals was to reduce capacity to or below the level of a mediocre winter season, and to cope with the spring problem by transferring resources from main shed weaving to pattern making for the next winter season. They implied that should a good winter season develop Gardiner would have to rely on getting weaving done on commission to cope with the domestic shortfall.

195

The workforce accepted these ideas. The alternative did not have to be explained to them, and the prospect of Gardiner of Selkirk bereft of it's weaving was even more distasteful to them than it was to the Board, for they would be bereft of their jobs as well.

These moves marked the nadir of Gardiner's operations, both in weaving and spinning. The winter 1983 weaving season developed sufficient strength for the commission weaving of 1460 pieces in April, May and June to have been contracted by mid March, and in the spinning mill, hitherto running two sets, three shifts, one shift was manned up to run all three sets. The year to January 1983 had ended with losses in both divisions totalling £54,419, a dividend repeated at 5p per share, but the operation in much better fettle than for some time.

A dividend of that order, worth £10,000 to Tootal, was not regarded as an adequate reward for their investment and while retaining their aloof posture, keeping their fingers out of the business of running Gardiner, our infrequent conversations with Norman Hornsby, and even less frequent ones with Alan Wagstaff, drove home their message that it was perhaps time to find a purchaser. Colin and I had both thought about early retirement, indeed I had intended to go when I was 55, in 1981, but at that time the pension situation was inadequate. But it seemed sensible, if the firm was to be sold, to arrange our affairs so that we could properly pensioned, go early, and to this end we arranged, for a one off payment for our pensions to be available in May 1986, when we would be 60 and 62 respectively. This made us more willing to promote the search for a purchaser.

The first moves were made by Norman Hornsby. Knowing Russell Smith, the Chairman of Allied Textiles, famous for having piles of money invested in Gilt Edged Stocks, he approached him and asked if he might be interested, and got a dusty answer. Allied had a portfolio of first class businesses which Gardiner could not match for quality.

The next attempt was made by me and Colin. Some years earlier, late at night in the bar at one of the industry conferences, Jock Mackenzie had talked casually of bringing Gardiner into his Scottish, English and European Textiles group (SEET), but at that time we weren't interested and let the conversation die. Now we thought we might revive the idea, and wrote to him seeking to revive the conversation.

SEET had, at that time, four members, all prosperous. First, the Mackenzie Harris Tweed business in Stornoway, run by the aptly named Harris Mackenzie. Then Peter Macarthur, tartan weavers in Hamilton, very prosperous and sharing Bill Rohe as agent in West Germany. Then Blackburns, flannel makers in Yorkshire and lastly Cree Mills, in Newton Stewart, who made very expensive mohair cloths, rugs and scarves, run by Bill Archibald. It was not inconceivable that our spinning and weaving operation might fit into that group.

The immediate response was encouraging. We were invited to London to talk the matter over, and after sending ahead the appropriate accounts and figures, made the journey. On 12 January 1983 we caught the British Caledonian flight from Newcastle to Heathrow and met Jock and Roland Leigh in their offices in London.

The morning discussions were of past results and future prospects, and of the way that Gardiner might fit into the group. Spinning was one of our hopes, but was not a rock solid selling point, for Harris Tweed yarn has to be spun in the islands of Harris and Lewis, and Peter Macarthur was a worsted based business, what woollen yarns they used being rather fine for our machinery. But things went well enough for Jock to take us to lunch at the Wig and Pen Club, opposite the Law Courts, where he did his very best to fill us full of drink to the point of indiscretion, failing marginally in the attempt. Lunch went on until late afternoon, by which time we had missed the last British Caledonian flight back to Newcastle and had a scramble to get to Gatwick to catch the last Dan Air flight home.

The outcome of that meeting was a visit by Harris Mackenzie, Alastair Bottomley, who ran Peter Macarthur, and Roland Leigh to Selkirk, where we showed them the mills and continued the talk about "fit". We agreed to make quotations for yarns which they used, and for finishing for Blackburns and in due course submitted them. By 10 March the SEET Board had decided that there was no "fit", and the idea was put into cold storage, with an invitation to revive it in 6 to 12 months if anything changed.

In the years 1983 to 1988 no less than eleven sets of conversations whose objective was the sale of Gardiner as a whole took place. Some were instigated by Tootal, some by Gardiner and some by the potential purchaser. Only one produced an unqualified offer of cash for the whole business. In July 1985 an obscure operator from Cheshire offered £650,000 for the whole business, a figure which seemed to us, with a first half profit of £177,000 tucked away and a net worth at the close of the previous year of £1,627,000, a pound or two short of reality. It is illustrative of Tootal's desire to sell that they suggested that if the prospective purchaser could be pushed up to £900,000 they would take £400,000 as their share. Happily that suggestion was never tested.

One of our major competitors in the spinning business, with strong knitwear connections, visited us as a result of Tootal approaches. We spent far too much time talking with them and disclosed far too much information for their proposal, when it materialised, was that they buy our spinning machinery for removal elsewhere and leave us with an empty building, and the weaving division. This we rejected without consulting Tootal and without asking the price. Yet another competitor with spinning capacity problems had a similar idea except that the machinery would stay where it was and the whole spinning and twisting plant would be bought. That competitor also had a weaving firm about the same size as Gardiner, and the proposal foundered when that weaver declined to join in and buy the Gardiner weaving division.

Perhaps the most bizarre proposal (from our point of view) was that put forward by our agent in Paris, Gigi Saada, who, with a Swiss backer, came across to Selkirk in 1986 to propose that they bought Gardiner for £840,000 on condition that Colin and I immediately bought back 30% for £240,000, with an option to take the holding up to 45% within 12 months at the same price. Selling at £2.10 a share and buying back at £2.00, putting only £90,000 apiece into our pockets, losing control and facing probably diabolical Capital Gains Tax problems didn't appeal to us, and neither the approach nor the proposal were ever reported to Tootal.

Between May and October of 1986 in depth conversations were held with Richards of Aberdeen, a firm prominent in the furnishing yarn trade, prosperous, and having a large lump of cash in hand from the sale of land to a supermarket chain. For a long time this looked like a deal which would come to fruition. Accountants analysed us and reported. Colin and I, and our new management team of Brian Roberts, Peter Harvey and David Stewart separately spent days in Edinburgh in deep and meaningful conversations, but at the end of the day Ray Dinsdale, the Managing Director advised his Board not to proceed, for reasons which he disclosed but later acknowledged to be unfounded. Whilst the deal looked good at the time, the best so far from every point of view, it was to our substantial advantage that it fell through.

In the winter of 1983/84, hearing through the grapevine that Heather Mill were going to upgrade their spinning facility, I wrote a half serious, half flippant letter to David Stevenson, head of the Edinburgh Woollen Mill, suggesting that buying Gardiner would be a much more cost effective option. That approach brought forth no response, but Heather's spinning was left alone. In April 1987 a serious approach by EWM was made to us which foundered, I believe, on the twin rocks of price and our refusal to disclose detail to what was a major customer unless a deal was virtually certain to be achieved.

Textiles, as I have said earlier, is a ladies tea party, Mary Rae, of George Rae & Sons had a son Ewen who, in 1987 was working for the Manchester Exchange Trust, a minor merchant Bank, and he persuaded George Kynoch, of G & G Kynoch, of Keith, to look at us. At the time Kynoch were determined to expand out of the trouble they had been in for years on the back of marginally improving results. Accounts were exchanged and a meeting arranged in MET's office in Edinburgh, a meeting which George Kynoch must have found supremely embarrassing, for at the time Gardiner, off a turnover twenty percent greater than Kynoch were making three times their profit and had a net worth half as much again as Kynoch. Kynoch's only positive feature was their stock exchange quotation, a valuable asset but no magic wand. That meeting was the last; Kynoch are now out of textile manufacture altogether.

All in all, from 1983 to 1988 we all put a lot of effort into attempts to achieve our joint objective of passing Gardiner on as a going concern, for cash; an objective ultimately achieved, but not until the end of this history.

The year which ended on 31 January 1984 saw a spinning profit of
£114,319 and a weaving loss of £23,943, a combined profit of £90,376.
Not too good, but an improvement on the previous year's loss,
celebrated by a further fifty percent increase in the dividend to 7.5
pence per share. The restructuring of the weaving operation led to
cold hard looks at other bits of the total operation. All unsold cloth
stocks were put on offer at whatever price they would fetch. A drive
was instituted through the agents to attack customers where price
was a truly insuperable barrier at list minus fifteen percent in an effort
to replace the lost EWM business, and in the spinning division a major
assault was made on "batch ends".

At the end of manufacture of a batch of yarn there is a quantity of
fibre which cannot be spun into yarn. In part it is the last of the fibre
coming off the cards, and in part it is sliver or yarn which has been
broken out of the production process and discarded. It is all perfectly
good fibre, dyed to shade and reusable when the next demand for
the same shade and count comes along, so the "batch end" is baled
and stored, awaiting that event.

The stock of batch ends has a relentless tendency to grow. Season by
season the shades in demand change, and batch ends in last season's
shades get stranded. Minor shades are yarn dyed rather than being
spun in colour, and a change of status can strand a batch end with
little hope of use. So a continuous watch on and analysis of batch
ends should be a prominent feature of spinning mill operations, but
it is never prominent enough. The back end of 1983 saw an overdue
attack on the problem, with lots with no immediate prospect of use
bulked together and dyed black or, where yarn qualities permitted,
lots originating from different yarn counts being merged into the
coarser count. All this costs money for dyeing, but reduces raw wool
intake and dead stock.

Against this depressing background, however, the Board felt cause
for optimism and set out on a number of positive courses. An agent
to sell yarn in Scotland, to intensify the efforts currently being made
by mill staff, was sought and ultimately Ronnie Allan, a sailboard
enthusiast from Ayrshire, was appointed. Against expected
improvements in the weaving business a yarn stock building
programme, ahead of the winter 1984 season, was laid out for

implementation. And, in the spinning division, changes in the way EWM, a major customer, operated, required yet more wool bins. Space for them was found in the end of the yarn store nearest to the spinning mill building and two bins each with 4½ tonnes capacity, complete with fans and pipework, were ordered to cost £22,000.

Gardiner were not alone in having a more than usual ration of problems to overcome. The majority of firms making up the membership of the three trade associations, The National Association of Scottish Woollen Manufacturers, the Scottish Woollen Trade Employers Association and the Scottish Woollen Trade Mark Association, were up against it to a greater or lesser degree. One of the products of the contraction of the industry was a proposal, rumbling along for some years but now to be taken seriously, that the associations merge. An idea not quite as simple as it might first seem, for whilst most firms belonged to all three, some were members of one association only for reasons which they regarded as sound.

But in 1984 it was clear that at the Conference, to be held at Turnberry, when the three Annual General Meetings would take place, the merger proposal would be put to the vote. The Gardiner Board decided to vote in favour provided that the more extreme proposal of including knitwear manufacturers and garment makers in the new body were dropped.

The Biennial Conference was generally regarded as a "jolly", but that of 1984 was to be significant in the affairs of the Industry and of Gardiner. As always the Conference had a theme supported by guest speakers prominent in their field. The theme this year was design, and the speakers were Barry Reid, of Austin Reid and George Young, Chairman of Jaeger.

Chapter Twenty Three

More Change in the Making 1984/1986

So, in March 1984, we all went off to Turnberry for the Conference. To get the outsiders view of design in the mid eighties, to eat and drink and socialise, to play golf and to merge the associations into one, to be called "The Scottish Woollen Industry". We were lucky with the weather and I believe it was one of the more successful conferences. Certainly Turnberry was, for many of us, the favourite venue, though it did not suit those from the far North. It was in this year, on the pitch and putt course while our wives were getting chamfered up for dinner that Edward Aglen, former Secretary of the associations and guest at the Conference, introduced me to the golfing term "NITBY". A depressing but all to often appropriate acronym. "Not in the bunker YET".

Over the years the designers and selectors employed by our customers had been getting younger and the percentage of females had been rising. The getting younger was real, not just as related to our own advancing years but when coupled with that produced an ever widening generation gap. The cloth buyer in the greatly reduced number of garment making firms, formerly and traditionally a man with vast experience of cloth of all types, knowing the identities of the prime producers of each and every type of fabric, had become almost extinct, to be replaced by teams of eager, well educated young people of limited experience.

It was significant, then, that both Barry Reid and George Young devoted a lot of time to the structure and composition of buying/designing teams at the present time, and both commented on the difficulties which they, as directors of those teams, had in coping with the generation gap, and on the communication problems which resulted from that gap. Clearly, if the generation gap was a problem within an organisation it must also raise difficulties in the supplier/

/customer relationship. Indeed it was a phenomenon which we had all, even young Brian, noticed, though none of us had been sufficiently impressed hitherto to think constructively about it. The 1984 Conference, and particularly the Reid and Young contribution were to change all that.

Returning to Selkirk the subject kept us occupied whilst opening the mail until it resulted in a discussion paper for the 20 April Board meeting. Appropriately headed "Thoughts for the Future", and having in its opening paragraph the cheering announcement that "IMJ and HCB are getting old, and so is GHB (Gerald Box)". The paper went on to discuss the generation gap, the future sale of the Company, the management structure required for the future and contained my declaration of intent to retire from active management two years hence in May 1986 whether the Company were sold or not.

The paper's proposal for tackling the generation gap was that we swiftly employ a young designer fresh out of the Scottish College of Textiles and immediately set about exposing "it" to the youngsters on the opposite side of the fence, taking it to sales meetings with customers, to trade shows and generally making sure that it became externally oriented rather than, as was the case with most of the older generation of designers, mill oriented. After very little debate, so strong had been the impact of the Conference, this was agreed. The interviews which ensued resulted in the appointment of Lynn Lawrie, in our view the best of a very good crop, to the post in the early summer. She turned out extremely well, and her departure for pastures new in 1987 was an unexpected blow, though we soon discovered that the quality of the students coming out of the SCOT remained just as high, and her replacement was soon found.

The debate on the future management structure was longer, for it hinged to a degree on who intended to retire and when. My position was clear and the idea of retiring in 1986 held some appeal for Colin, so the discussion went forward on the basis that the new management team would have to be in place and functioning by that date.

The nature of my successor provoked little argument; a Chartered Accountant with suitable industrial experience, to be in place by January 1986 at the latest. Brian's experience was management oriented. Though he had done a stint of selling full time at L. J. Booth, he had had no design experience, LJB being makers of piece dyed fabrics, which made it quite clear that the other member should be a design/sales director. Noting that time would be needed to draw up the detailed structure of the new management, and the job specifications which would flow from it, the Board concluded that, if the design director was to be in place by mid 1985, work on the specifications must be carried through without delay, and the search should begin later in 1984, for the successful candidate might have to give substantial notice.

Also on the agenda for that meeting was the subject of a new computer within the next two years. The System 10 machine now had a successor, the System 25, which would do everything very much faster, and which had the added advantage that the existing suite of programmes could be machine translated to operate the new machine. But in the interim the current programmes should be modified in the light of the experience we had of running them, and the Board agreed to employ a SCOT trained man, qualified to HND level, to oversee the programme changes and the move to a new machine, recognising that the post would probably have a limited life having regard to the extent to which modern systems are "user driven".

The clear determination to continue weaving, for without that there would be no need for new designers and design directors, prompted the meeting to authorise a 500 copy print run of the booklet which had been brought out to tell knitwear customers about Gardiner, but expanded to extol Gardiners weaving and woven product as well.

Now that we were to all intents and purposes committed to retirement in 1986 Colin and I began to take an even more serious interest in our pension situation. We had left the Company's staff scheme and each now had a personal scheme, designed to generate a certain fund by a certain date. Periodically information came from our brokers setting out what the pension would be at varying funding rates and the variations were sufficiently large to cause us concern, for if, on retirement, the rate was adverse to us our pensions would suffer.

In charge of pensions at Bowring Macalaster and Alison Ltd. was a young man named Patrick Wynne, who seemed to know rather more about pensions than anyone with whom we had hitherto dealt, so we took the problem to him. Was it not possible to "nail down" the funding rate so as to yield pensions of two thirds of our final salaries, whose size was by now known to close tolerances, for sure? Certainly, replied Patrick, coming back to the subject a few weeks later with the news that our needs were met, and the saving to Gardiner in Pension costs over the next two years would be £80,000. Was that worth a one off fee of £2,500? We thought it was.

Throughout 1984 trade improved as the recession abated, though not to the extent that a return to the former scale of our weaving operations could be contemplated. The new format of mixed main shed and pattern weaving was working very well. On the machinery front we decided to replace the Gilbos cone winding machinery in the spinning mill with two, 16 spindle, Swiss, Scharer machines at a cost of £100,000 which, running three shifts would, according to their makers, cope with the whole production. Though the production calculations leading to that conclusion followed trials of winding our yarn we were soon to find that forecast and reality were miles apart. After the event the flaws in the calculations seemed obvious and we were to enter upon a long running battle with Scharer about compensation for failure to perform. To add to our woes we failed to find a buyer for the redundant Gilbos machinery which, since it was wholly written off was more of an embarrassment in terms of space than of cash.

In the Autumn of 1984 the Scottish College of Textiles held a two day seminar and exhibition of computers. I visited it, for we had not yet contracted for a System 25 machine, and looked over what was on offer, talked to all and sundry about the capabilities of the various machines and met Peter Watson of Berkeley Computer Services, of Glasgow. In no time at all we found ourselves on the same wavelength and that first conversation, in which he outlined the user driven and multi programme capabilities of the Honeywell DPS 6 machine, led to later meetings in which we explored the possibility of writing a new suite of programmes for Gardiner to run on that machine, and developed that into the idea of writing a system which would be saleable by Berkeley to the weaving and spinning trade at large, with cost advantages for Gardiner. Berkeley's "Masterpiece" textile system and Gardiner's 1986 Honeywell DPS 6 installation were the fruit of these conversations. The project, costing £50,000, was approved in

February with implementation planned for the summer holiday on hardware to be delivered early in April to allow truly comprehensive testing of the programmes as they were delivered.

Also in the computer field the woven fabric design programme being developed by the Scottish College of Textiles, later to be christened "Scotweave", was researched by our designers. Their conclusion at that time was that the system was not yet well enough developed to make it worth acquiring, but that it should be kept under review.

The improvement in the weaving business which had prompted a fair volume of commission weaving in the early part of 1984 set us thinking of multi shift weaving again with a target of 15,000 pieces a year. The idea, coupled to a substantial reduction of the price of the product which would be made possible by the increased output, was put to our agents who all agreed that the price change envisaged would not bring in the required volume and the proposal got no further. Three shift working in twisting was discussed and agreed, with reduced manning and the sale or scrapping of one Boyd machine, for introduction when the new Scharer winder was in service. The logic behind this was the rising demand for twisting flowing from increasing sales of two fold Shetland yarn to the knitwear industry, coupled with the reduced workload on the twisting operatives which would result from the substantially greater weight within the same dimensions of a Scharer "precision wound" package, as compared with that from the Gilbos machines.

On the cloth sales front our agents in all our important markets were causing great concern. In London where, following Ken Smith's retirement several years earlier Gerald Box had been soldiering on by himself, his activity was greatly reduced by illness. His lungs had given trouble for years and it had become serious enough to require operations and, ultimately, the removal of one of them. All this meant that Gerald simply had not the strength to carry on alone and we had to explore all sorts of propositions from Gerald taking a partner to opening a London office jointly with a German firm which was also represented by Gerald. The chosen partner backed out, the Joint Office failed to materialise and ultimately the agency was handed to an established firm, S. Frank Cook & Co., who came to an arrangement with Gerald to use what services he could still provide.

In Germany Bill Rohe was terminally ill with cancer with Ingo Monkemuller taking more and more of the load. Though Ingo had been with Bill for years and was a first class salesman he did not have Bill's entrepreneurial flair, so necessary in a business which, besides selling cloth on commission, also bought fabric and merchanted it. But there was no obvious alternative to backing Ingo in his conduct of the Rohe business. In New York, where we had been represented by Tony Dawson working out of the Dick and Goldsmith office for years the problem was the same, the terminal illness of Tony. Here we had, theoretically, an existing set up to carry on with, but that organisation failed to employ a suitable salesman which led to a period of ineffectual acrimony which culminated in D & G instituting proceedings against Gardiner when, in June 1986, we appointed Eighteenth International to represent us.

While all these problems exercised the Board the search for the new management team began. Years ago we had used a "Head Hunter", one W. Thomson, in the search for a designer, and we asked him to seek out a design and sales director by a combination of advertising and head hunting. It might be thought that the contraction of the industry which had occurred over the past few years would have left the ground littered with likely candidates, but this was not the case. Those "dropped out through the bottom" tended, if young enough, to leave the industry for pastures new, whilst those of appropriate age simply retired. So all those approached directly, and all those who replied to the advertisements, were in active employment.

As is always the case in senior appointments such as this, three quarters of the applications could be discarded at the first reading and, after detailed consideration of the rest and after discussing them in detail with Thomson we selected four to see. The interviewing was done at the Crest Hotel in Carlisle, for only one applicant was based in Scotland, and after seeing them all Colin and I said to each other that if Peter Harvey, our first choice, or Mr. X, our second, did not come to terms with us we would be unable to retire unless further search revealed a hitherto unknown acceptable candidate.

In the event Peter Harvey, then doing the same job at Robert Noble, in Peebles, part of the Dawson Group, agreed to join us and arranged to start on 1 July 1985. This was a very successful appointment for

Peter's style, and his ideas of product and marketing, were entirely different from ours and he swept in like a breath of fresh air at the time when it was most needed. The philosophies which had underlain our prosperity over the years had run their course and Peter came aboard with just the right new ones.

The search for the Finance Director was to be conducted through our Auditors, Deloitte, Haskins and Sells, and was to be left until the Autumn.

The year to 31 January 1985 produced a spinning profit of £80,382, and a greatly reduced weaving loss of £5,670 to give a combined profit of £74,622. We were going in the wrong direction again, but the drastic reorganisation of the weaving division was bearing fruit. The first half profit in that division had risen from fifty seven to eighty three thousand pounds, whilst the second half loss had likewise moved upwards from eighty three thousand to eighty nine thousand pounds. Little wonder that we would happily have placed large sums in a Swiss bank account for anyone who could come up with a permanent solution to the perennial spring season problem.

Though we could not know it at the time the actions which we had taken from the time of reorganising the weaving down to the appointment of Peter Harvey were, together, to allow us to retire contentedly, to set Gardiner on the road to record profits and make possible its ultimate sale.

Chapter Twenty Four
Winding Down and Handing Over 1985/1986

With the appointment of Peter Harvey the end of the thirty year reign of Brown and Jackson could be said to have been entered into the programme. Clearly the transfer of management functions would begin quite soon after his arrival on 1 July 1985. A man of his experience and drive could not be expected to remain in the shadows until the incumbents departed the scene. Meanwhile the conduct of the ongoing business needed no less attention than hitherto.

Prominent in the problems area was the new Scharer winding machinery, which was not only not performing to specification but, on the evidence by now to hand, could never cope with the output of the spinning plant. Re-worked production figures, based on the experience we now had of operating the machines, showed that the minimum modification needed was four more spindles on each machine, an increase of 25% - the measure of the errors in the maker's earlier calculations. Gardiner thought that they should be provided free of charge but, notwithstanding the validity of our arguments, we could do no better than to have them offered for £10,000 installed and running complete with clearers and splicers, with an empty package conveyor and modifications to the input package holders on all spindles thrown in. Gardiner's suggestion that the makers re-purchased the machines and compensated us for all our expenses did not get very far down the track. Nor was it really desirable, for the Scharer did wind a better package, and would bring economies in following processes.

Peter Harvey's first request for capital equipment, made even before he arrived, was for a power guillotine for cutting patterns, and this was put on order the week of his joining, at a projected cost of £5,750.

By June the computer programming had fallen behind schedule to the extent that summer holiday implementation now seemed unlikely, though that target was not yet discarded.

Among the ideas mooted at the June Board Meeting was the suggestion that we have skirts made up on commission from Gardiner fabrics which would match or tone with knitwear made from our yarns, and offer them to the relevant knitters. Whilst not ruled right out the plan did not generate much enthusiasm, partly because of the added strain on our design department, and partly because of the old fear of competing with our existing skirt making customers. It was the case that where such a demand really existed, as in the British Home Stores/Dukes and Marcus operation, it was already being met by the sale of cloth as well as knitwear yarn.

July saw two Board meetings, and August one, the increased activity pointedly marking the end of the "do it over the morning mail" era, formalising the new management structure and setting the pattern of Peter's early activities. A journey to North America in September was planned, followed either directly or in October by a visit to Japan, markets in which Peter had developed special relationships which must be nurtured. At the second July meeting, on the seventeenth, the antiquity of Gardiner's office and showroom facilities and decor were drawn to the attention of the retiring management, and a reallocation of accommodation was proposed, with agreement that we take professional advice on the decor. When you live in accommodation for as long as we had you don't see it getting out of date. It struck Peter forcibly right away.

The other area needing attention was sample making and pattern cutting, with particular reference to the number of bunches to be cut from each range and their shape, function and distribution. The immediate outcome of these discussions was putting the sample warping machine on double day shifts and beefing up the pattern room staff, with a further drawing/knotting operative to be sought urgently.

That meeting closed with the offer to Brian Roberts, and his acceptance of, the post of Managing Director when the old men retired.

The first half, ending at end July 1985, showed a continuation of the progress towards true recovery. Spinning profits advanced from thirteen to forty four thousand pounds, while weaving went from eighty three to one hundred and thirty four thousand, a near doubling of first half result from ninety six to one hundred and seventy eight thousand pounds. A good foundation for the new team to work from, but not worthy of mention in the minutes of the 20 August meeting which was more concerned with driving onwards to the future. The various agency difficulties mentioned in the previous chapter were reviewed and steps to the ultimate conclusions initiated. And, importantly, a possible Japanese representative, Mr. Uchinuma, who had been recommended to us, was coming to Britain, and would visit Selkirk so that we could weigh each other up with a view to his appointment.

And, significantly, Peter Harvey's personal, sustained attack on the spring season problem began with the decision to add five, spring oriented, bought in, yarn qualities to the string; final recognition that our woollen spun yarns on their own would not even make a dent in the sales shortfall in that season.

Again significantly changes in the price structure for West Germany designed to allow us to make money out of shorter runs were initiated. The old basis of a list based on six pieces per colour, with a small surcharge for threes, which had been the foundation of "High Straight" was formally abandoned. Change was certainly in the air.

As 1985 rolled into Autumn the commitment to the change of management grew yet more solid with the appointment of David Stewart to be the future Finance Director. Brian Roberts and I, with Jeremy Burnett, "our" Deloitte partner, had picked him from the four short listed candidates we had interviewed. An ex Deloittes Chartered Accountant, he had industrial experience in the frozen food industry, one which, we learned to our surprise was, like our own, affected by the problems associated with the water content of the product. David joined us on 1 December 1985 and his arrival forced us, for the first time, to formalise our Company car situation, setting parameters for who could have what.

Hitherto the cars had been on a strictly "need to have" basis , or so we tried to persuade ourselves. For many years there had been but two, one each for the executive directors. Recruiting Murray Tait, the designer and David Knox, the spinning mill manager for the North of Scotland, where they had each had a car on a less than essential need basis forced us into providing two cars to which the criterion of absolute need did not attach. Naturally, this caused some discontent but, with the addition of Brian Robert's car, which was a need, the fleet stabilised at five, with choice of car fairly informal but for Murray Tait's and David Knox's which were strictly Ford Sierra or equivalent.

When Peter Harvey arrived he chose for himself the top of the range Sierra Ghia, so a similar car was offered to David Stewart, but he had a feeling for a Saab, costing no more, so he had one of those. The fleet was now seven and, having only one van for mill use, we increased it to eight by adding a Ford Cortina estate for use as a spare car and small van. Breaking away from the old, cosy, way of doing things was bringing changes we hadn't foreseen.

Peter Harvey agreed with the philosophy of selling through agents but he did increase, substantially, the amount of travelling, particularly overseas, undertaken by himself and the designers in support of the agents. Where he did depart from past policy was in involvement with trade fairs, which we had, hitherto, largely eschewed, believing that the product was better sold by knocking on doors. In the Autumn of 1985 Peter had persuaded the Board that we should show at a mens wear fabric show held at the Dorchester twice a year under the auspices of the IWS and the SWPC, and further, that as soon as we could gain admission to an already fully subscribed event, we should take part in the Premier Vision show which was held twice a year in Paris. It took Gardiner some time to gain admission, but it was achieved in 1987 after continuous lobbying by Peter Harvey. The projected annual cost of these two shows was put at £20,000.

Product variety was broadened by the introduction of fifteen shades of a Donegal yarn made in our own mill with a view to capturing a share of the substantial stream of business in Donegal type fabrics which went mainly to mills in Northern Ireland.

On a lighter note, the tenth anniversary of the opening of the spinning mill was approaching. In December 1975 when its first batch of yarn had been made the weaving division had been on short time and we cancelled the party we had planned as inappropriate and the new venture quietly crept into life. Now, ten years on, with the venture established and the Company as a whole prospering we felt that we should mark the occasion and decided to give each employee of both divisions a package containing a bottle of Scotch and a bottle of Champagne, to be kept secret till distribution time. Despite the communicative skills of the local grapevine we did achieve this and no hint of it got out in advance. In an organisation working in its various departments day work, double day shift and three shifts the issue of the parcels was akin to a military operation. The obvious parameter was that no-one should get his or her parcel until they were actually leaving the mill for the day. This involved managers and foremen being on hand to make the issue at two in the afternoon, four thirty, ten in the evening and six the following morning. All this was achieved, and the stocks awaiting distribution were kept secure. Apart from the annual mill dance, which had become an established event, that was our first official celebration since the Centenary in 1967.

The oldest of our Sulzer weaving machines was now eighteen years old, which did not mean that it was worn out or anything like it, but there had been a lot of development in weaving machinery in that period. Though water jet and air jet looms were not appropriate to our woollen trade the various rapier looms which had been introduced since 1967 certainly were, and whilst the Sulzer was still in the forefront of weaving technology we felt that we should explore the relative suitability of Sulzer and the rapier machines. There were three machines that we felt should be examined; the Dornier, a rigid rapier machine, and the most expensive. The Sometmaster, the latest development of the most widely used of the rapiers, and the Vamatex, a machine developed by the inventor of the Somet when he parted company with that firm, said to be the fastest of them all.

The rapier machines all ran faster than our Sulzers. Sheer mechanical speed is of little use unless the yarns used are capable of taking the strains of acceleration involved in these high speeds, and rapiers were said to be kinder to yarn than the Sulzer "bullet". Additionally all of the rapiers could put eight weft colours across an infinite number of warp colours, whereas all but six of our Sulzers had a four colour weft capability, the other six being six colour machines. The Board decided to research the theoretical and financial aspects of changing to one of the three above mentioned rapier machines, and to conduct weaving trials on them all. All this would take time, but nothing lasts forever and it would be as well to be prepared and informed. By April 1986 the paper research had narrowed the field to Sometmaster and Vamatex and the weaving trials were restricted to those machines, though during that year we were tempted to add the Picanol machine to the list, and trials were conducted on it as well.

The limited weaving trials which can be done on other peoples machinery seldom lead to data sufficiently compelling to inspire a decision and, having done these trials the Board's conclusion was that Gardiner should take up the offers to instal and run at Selkirk a Vamatex machine and a Picanol Machine. In the event only the Vamatex was delivered, early in 1987. By July 1987 the trials were complete and the machine was removed. We now had all the data we needed to compare the Sulzer and Vamatex capabilities.

The crucial influence on weaving efficiency is the frequency of breaks in the warp and weft yarns leading to stoppage of the machine. It controls not only the efficiency of the weaving shed, but also the number of looms per weaver with all the impact on costs which flows from that. We had accumulated a vast bank of data from the trials which we sent to Management Services Centre for analysis, along with a similar block of data derived from a parallel study of Sulzer weaving. The analysis duly arrived, seemed inconclusive or even favourable to Sulzer in certain crucial cloths, was debated and by August 1987 the Board had decided not to buy Vamatex, but to continue to think about weaving machinery.

Textile arithmetic done by other people, whether they be selling machinery or just doing a study to a brief, must always be looked at from the front, the side and through the paper from the back, and this report was no exception. It was not until the agent for Vamatex, the redoubtable Harvey Wilson, bluntly told me that he didn't believe that our break frequency study could have been properly conducted that I, by then retired, sat down and really analysed it and thought about it, and discovered that it was all based on simple averages, not weighted averages taking account of the size of each study. Reworking showed a clear advantage to Vamatex, so the debate was reopened and was still raging until silenced by the sale of Gardiner to a loyal Sulzer user.

Prospects for the weaving division were, by December 1985, good enough for a move back to shift working in the weaving shed, and night shift working was begun on 16 Sulzers and three Hattersleys. At the same time the darners bonus arrangements, which had drifted right out of line, were altered to bring them back into line, a change which could not have been achieved in the Wilson years.

Programme testing for the new computer had been intensive throughout the second half of the year and by December was far enough advanced for early February 1985 implementation to be contemplated. Because loading the sales side of operations into the machine needed every sales order in the book to be entered so as to generate all the yarn requirements, capacity charts and the like which the system produced, it was important that it be done before the winter season orders made it too big a task. The total integration of all parts of the system demanded that setting up the status quo in the files had to be done all at once before the system could be used.

In the event the implementation was carried out with only one major hitch over the weekend which included the year end. The computer staff, Andrew Jackson and Deborah, with Margaret, who had run the previous 1901 machine, came in over the weekend to do the tedious keyboard work, whilst the sample room team of Alec Kinghorn and Bill Martin came in to monitor and collate the output of order confirmations, piece tickets, progress tickets and yarn issue slips, with Tom Spence the yarn store manager and Ian Cooper the weaving manager there to test check and file for use their share of these documents. Steven Pratt, from Berkeley, came over for the weekend as insurance lest anything should go wrong. He had written most of the programmes.

Saturday went well and closed with a fine pile of output, all of which appeared to be exactly right, and about seven in the evening the system was closed down for a nine o'clock start on Sunday. That day we decided to operate in batches of seventy orders between print runs, a decision which turned out to be wise, for when we came to print the first batch we got all of yesterday's product out of the printer again as well. All yesterday's yarn requirements and capacity needs had been duplicated and the implementation was heading for disaster. Steven Pratt proved his worth there and then by writing a small programme which reversed the duplications, and modified the main programme to prevent further duplication. Without him on hand the implementation would have to have been abandoned.

By Sunday night the back of the job was broken. All the orders due for delivery on dates requiring action in the next two weeks were in system, and the rest could go in in ordinary time along with current input. For the first time in three computers we had not had to back off and start again months later. It had a lot to do with programme testing.

The year end came up on 31 January 1986 and in mid February we learned that spinning had made £174,569, weaving had made it's first profit in four years, £3,070, a total pre tax profit of £177,639. An interim dividend of 10 pence a share, payable in April, was declared.

When David Stewart was reasonably settled in I moved out of my office and into the room behind the showroom, until recently the home of Joe Jones, who did our costings and chased our progress until he retired in 1985. It was the room which I had occupied when I first came to Selkirk in 1954, and so a suitable place in which to see out my time. Windowless and cheerless I had been glad to move out of it in 1956. Peter Harvey had opted to work in the showroom, also windowless but roof lit and much more spacious. Colin stayed where he was, in the room at the front which was to become the new showroom.

Consultation between Peter and Michael Vee produced a design for refurbishing the reception area and the old and new showrooms. All dark panelling, framed prints and brass light fittings. Very Baronial Scottish, a great improvement on what it replaced, but known to some as the funeral parlour. Fortunately when, in 1987, a plumber working on the valleys in the office block roof left his blow-torch going over lunchtime and set the roof on fire the damage, which was considerable, left the parlour unscathed. In rebuilding we did what we should have done in the first place and improved and reallocated space in the whole of the office block. Moving the toilets into what had been my back office made space for two new offices with windows and allowed a general environmental improvement. It was fortunate that no-one was hurt, and that no damage was sustained by the office equipment or the computer, and no records were lost. The plumber was not popular, either with his employer or with Gardiners.

Starting in March 1986 a pattern of monthly Board meetings was established with a view to keeping Brown and Jackson closely involved with operations, but in a "hands off" mode. At the March meeting Frank Miller, a yarn agent, was appointed to sell weaving yarns in the North of England. The Board felt that other firms would make such different use of our standard yarns that no harm could come to us, and increased use would lead to bigger batches and greater stock turn. And the Scharer affair rumbled on, with a decision to seek legal advice on the possibility of action to recover some of the costs which the disaster had imposed on us. I resigned from the office of Company Secretary, effective 17 March, David Stewart being appointed in my place.

The last Board meeting which Colin and I attended as executives was on 27 April 1986, when Brian Roberts was appointed Managing Director from 12 May, the first such since 1956. The market for Shetland yarn was in a gloomy state and spinning cutbacks were mooted but not decided on. A major claim from Campsie Knitwear was reported to have been settled for £7,260, less what we could get for the faulty garments. And finally, probably stemming from the management changes, Tootal sought, and management agreed to write, a paper covering the policies and prospects for the next two years.

In 1986 the 12 May, my birthday, was a Monday. We both stayed at home in the morning, and at lunchtime Colin and Mary took me and Joan to "La Potinere", in Gullane, for a birthday lunch. A change from working. A big change, after thirty two years and four months and thirty one years and seven months respectively.

Chapter Twenty Five
Prosperity 1986 to 1988

Gardiner, in common with the majority of firms in the wool textile industry, now embarked on one of those parts of the cycle when things go well and the temptation to believe that it can never go wrong is hard to resist. Cycle apart, the change of direction of the weaving division instituted by Peter Harvey began to bear fruit in a spectacular way, and the prosperity of the knitwear industry brought the 1980 expansion of the spinning division to life. But even in good times life in wool textiles is never easy, and as new ventures moved forward the last remnants of "High Straight" were looking distinctly frayed at the edges.

Marks and Spencer had ceased to use our cloth for the childrens coat, and with that what had been a vast core business for Gardiner disappeared. The Canadian trade was changing, both in the customers with whom Gardiner dealt and in the nature of the orders coming forward, but competition there was fierce, with Saddleworth, Abraham Moon and Woolly Mill all fighting desperately for the orders being bid for by Gardiner. In winter seasons our share was rising, but in June 1986 Peter reported that there would be no spring order for Gardiner. Saddleworth had scooped the pool with a new fabric at a price which represented an offer which no buyer could turn down. At home we secured the crucial Burberry contract in June, but at last year's price and in slightly smaller quantity than we had sought. The difficulties experienced in hanging on to the last of our true bulk customers were illustrated in August, by which time the contract was not yet on paper, and again in September, by which time the contract was signed, but no colouring instructions had yet materialised. The contract's virtue, given the razor sharp price, had always been that a large part of it could be woven in the slack period from September to January; crammed into the peak part of the year it would cause chaos, indeed that is how we had lost half the contract in an earlier year. So the Board agreed to guess the colourings of three quarters of the contract and start weaving. By December the Board, with the Winter 1987 season developing nicely, worried that the weaving of the contract was not going fast enough and might lead to peak season problems!

The spinning division, in the spring of 1986, was coping with a depressed market for Shetland yarn and by June was reduced again to two sets working three shifts, with 16 redundancies. Indeed, by August the whole mill bar the weaving shed, benefiting from its new format, was on short time. But the corner was just out of sight and getting nearer by the day. By mid 1987 spinning was not only back on three sets, three shifts but the cost of increasing the size of the plant by one further set and two frames, with appropriate winding machinery, was being investigated. When, in December 1987, it was reported to be a million pounds, the Board, mindful of the fickle nature of the trade, opted for the status quo even though the year had opened with nearly six hundred thousand pounds in the Bank.

The improvement in the spinning division did not mean that it was problem free. The trend to finer, softer yarns was a fact which could not be ignored, and as early as May 1986, against the background of Rennies of Mill of Aden having introduced a "soft shetland", yarn Gardiner set out to design one, moving it from the standard eights count to tens, some twenty five percent finer. By August the composition of the new yarn had been settled after numerous trials, and after introduction as part of the range it was promoted, in March 1987, to the status of a stock service product in 24 shades.

Nor were fineness and softness the only ongoing problems. When building the spinning mill we had opted for production of yarn with a low oil content, partly to reduce fettling problems and partly to take advantage of resultant savings in cloth finishing costs. When the sale yarn activity, principally to knitwear manufacturers, developed, we struggled to convince our customers that our slightly higher price was mathematically justified by increased garment yields per kilo of yarn, and in the main we succeeded. But from time to time we encountered arguments that knitters finishing routines had to be changed to use our yarn, and when that led to some customers dropping us because of the problems created by running our yarn in tandem with that of other suppliers we had to think again, and in December 1986 we surrendered, and increased the oil content of the Shetland yarns from 3% to 8%. The change over was difficult, for customers had to be quoted two prices until it was completed and stocks of three percent yarn used up, and the count of the yarn in its new, oily, form had to be changed in the computer in all cloth patterns which used the shetland qualities. An immediate, capital, consequence of this change was the need for greater spinning oil tankage, which was ordered immediately. The new, L10 supersoft shetland yarn was oily from the start. It is ironic that what nearly prevented the building of the spinning mill was our own early lack of perception of the effect of low oil on price, count and yield.

The involvement of Brown and Jackson in retirement was rather more than that of disinterested non executive directors. Quite apart from remaining substantial shareholders and attending monthly Board meetings each devoted time to specialist contributions to the management of the Company. Colin Brown continued as a member of the Export Corporation and helped with the organisation of Gardiner's presence at trade fairs, and attended the fairs, doing his stint on the stand, always a wearing task. As Gardiner moved from only showing at Fabrex and the Dorchester to involvement with Premier Vision in Paris it was an increasing, not declining, workload. I continued to monitor the new computer programmes, and liaise with Berkeley in the correction of the faults which inevitably came to light. Indeed it was only because of my involvement in setting up the system at a firm of weavers in Northern Ireland that a basic error in the calculation of batch yields incorporated in the programme and not found by our testing came to light. Beyond that I developed specifications for programmes which would be useful to management and increase the usage of the hardware. Cloth and yarn costing were early examples, whilst the research into Hamel twisting, where yarn could be twisted to a pre determined length, prompted specification of a programme to incorporate length calculation in the production documents of all cloths containing twisted yarn. This led to the knowledge that, as our designers did not "centre" patterns when writing tickets, leaving that for the warper to do, every yarn weight calculation performed by the computer was wrong. Only to a very minor degree in ninety nine percent of designs, but wrong none the less, so another specification was developed to have the computer centre the pattern before writing the production documents and calculating the yarn requirements. The writing of chainmaking instructions for use on the Hattersley, pattern making looms, was a time consuming task for the weaving manager, and a specification was written for that, with a view to having the programme developed by the Scottish College ofTextiles to run on any IBM compatible personal computer for possible sale throughout the industry. Except for the batch yield case none of these programmes had been written by the time I left the Company.

With this heavy workload, and the preservation of fees for the other Directors in mind Colin and I entered into consultancy contracts with the Company!

Jack Donovan, our agent for yarn sales in the United States, had achieved little but losses on an abortive stock yarn operation, and his contract was terminated and he was replaced in February 1987 by The Yarn Sales Corporation of New York. Laidlaw and Fairgrieve were well entrenched in North America and we and our new agents knew that taking a bite out of their market would not be easy. Meanwhile, driven by Dr. Tessari and George Grange, Italy was becoming an important market for yarn and soon became our biggest export market by far. At home on the cloth side, Gerald Box made a brief recovery from illness, but by March 1987 was not well enough to continue, and retired.

The Vamatex and Sometmaster weaving trials, having produced results deemed unpropitious, flared into life again when the averaging errors were discovered and corrected, but were shelved in the summer of 1988 when negotiations with the Sulzer dedicated S. Jerome & Sons Ltd., were in progress. However, other departments of the mill used machinery which was past its best. The wet finishing department put most of the product through combined milling and scouring machines which had been bought second hand many years ago and in August 1986 the investigation of replacement machinery began. Various equipment was considered until, in December 1987, the Board decided to buy two Holmes Heaton two channel combines for £58,100 each, with associated installation and works to cost a further £12,500, for commissioning at the summer holiday in 1988. This was achieved with the machinery operational when the mill reopened.

The entirely manual darning or mending operation had, with the introduction of Sulzer weaving, been provided with new work-tables and stances designed to accept the plastic tube on which the cloth was rolled, eliminating the need to take it off the roll to make it manageable. Now, with new standards of quality and new techniques being applied the idea of using mechanised tables to pull the cloth across the darner's field of view, hitherto a manual task, was mooted. This would allow the whole of the operative's effort to be devoted to actual mending. The debate of this was still going on in October 1988.

The more the benefits of measuring the length of yarn, rather than just weighing it and relying on the count/weight ratio to provide the correct length were looked into, the more attractive the idea seemed. Using weight it was always prudent to oversupply rather than the reverse, and this resulted in a myriad small packages with just enough yarn left on them to make them worth re-winding, at great cost, being returned from production. These could either be kept identified by batch for use in the most appropriate circumstance, or issued at random with the constant danger of mismatching. Another added cost was the "runner making" involved before warping, that is winding bigger packages down to smaller to provide the number of cones needed by the warper, by its very nature an expensive and hit and miss wind. If all yarn winding could be measured many of these costs and dangers would disappear.

The twisting plant was aged, leased and well into its secondary leasing period and our researches into the types of machinery available to replace it included the Hamel system in which the yarn was parallel wound before being twisted at high speed. The winding process included accurate, pre-settable measuring and this feature, among the other benefits offered by the machinery, caused us in May 1987 to accept Hamel's offer of a small unit on trial. The tests were successful and in June 1988 Gardiners ordered 40 pre twisting and 180 uptwisting spindles from that firm, the third complete re-equipment of the twisting department in thirty years.

Elsewhere the capital equipment programme called for two more wool bins at a cost of £26,000 and 16 weft accumulator sets for the weaving machines for £25,000, a Mahlo weft straightener for the tenter, £28,000, and two beam lifters for £6,000, all of this approved at the August 1987 Board meeting. No wonder the capital spend for 1986/87 was £28,994, rising to £127,728 in the following year, all preceded by £170,434 in 1985/86. Also on the capital front the Scharer claim for extra cost resulting from incorrect manning forecasts over the life of the machinery was quantified at £230,000. Half hearted efforts were made through our lawyers to recover at least something, but the claim was ultimately dropped.

Apart from computers for data processing there were two applications special to the textile industry in which Gardiner took a continuing interest. The first, already mentioned, was the weaving design programme "Scotweave", which we were keeping under review. By May 1988 the programme was sufficiently developed for Peter Harvey to say that he was seriously interested, and by early 1989 it was installed in his office in Selkirk, with PJH wondering what he had ever done without it! The other system on offer was for the wool blending process.

Almost all batches of coloured wool contain a number of colours of fibre which, when blended together, make the shade required. Even batches which COULD be made from one dyeing seldom are, for the bulk dyeing will almost always be sufficiently off shade to need a touch of something else to "pull" it on shade. All this combining of colours to achieve, exactly, a predetermined result is an immensely skilled task carried out by an expert usually known simply as "the shader". Learnt by experience it requires a special eye for colour and is a very time consuming occupation.

Coming into use in the industry was a computer programme which could analyse a required shade, and, from a library of component shades already in its memory, printout a series of recipes from which the shade could be made up, also indicating the cheapest formula in terms of dyecost. Such a machine could not only specify the recipe for new shades, but could reconstruct old shades to achieve a lower dyecost. And when the business of putting an "off" shade right arose it could compare the standard with the batch, and indicate what to do to put it right at least cost. It all sounded too good to be true, and I was despatched to Huddersfield to see the system in operation at Fred Lawton's mill.

There seemed little doubt that the system worked so, in December 1987, we ordered it at a price, hardware and software, of £35,000. The kit arrived in February 1988 and after heroic efforts by Colin Ford setting up its memories and standards the system was operational by June 1988.

Whilst all this was going on the year to 31 January 1987 ended with a spinning profit of £105,514, a weaving profit of £53,291 making £158,805 in all, from which a dividend of 10p per share was declared.

For many years Gardiner had had a representative on the Board of Governors of the Scottish College of Textiles. I had held the post, nominated by the Scottish Woollen Industry and its predecessor Association. For eight years I had been Chairman, standing down in 1984 to provide a change of blood. In my later years as Chairman the College put up a proposal for a Technical Centre which was, in my view, pure empire building and anathema to me, but was beloved of the Scottish Woollen Industry. The scheme was liked by the SWI but even so I consistently opposed it. The Council of the SWI dropped me from my Governor's appointment the day I retired, and appointed Brian Roberts in my place. What a pity, for now I had time to be useful. Notwithstanding all that, Gardiners links with the College were solid and when Peter Harvey proposed that we sponsor a design competition for the students there was only enthusiasm. The competition was called "Natural Creativity", and required designs to be made using only naturally coloured wool, quite a stiff test for the students.

The Chairmanship of the Company had been on a rotating, meeting by meeting, basis for thirty years, during much of which the shareholding agreement's demand for an independent had been quietly ignored. That clause had been cancelled, but we all felt that an independent would be a good idea. But meanwhile, in February 1987, Colin Brown was appointed Chairman while we thought about a suitable candidate. In August 1987 the Board decided to offer the Chairmanship to Jack Shaw, who had conducted the first audit in which I was involved as an examinee, and who had recently retired from the Edinburgh Managing Partnership of Deloitte, Haskins and Sells to become Chief Executive of a new venture, Scottish Financial Enterprise. Jack, being new to the SFE, itself a new organisation, needed time to see if he could undertake the commitment. Before he said yes (or no) the negotiations with Jerome were under way and the offer was withdrawn.

New blood churned out new ideas and challenged long established practices and firmly held beliefs. Sometimes the old ideas were vindicated and sometimes not. One practice challenged was that of making pattern ranges and sample lengths on conventional looms, on the ground that the weaving speed benefit were they made on Sulzer machinery would be outweighed by the longer setting up time of that machine. The old view was, in the end, held to be correct, but the changes in the nature of the pattern making operation being forced by the altered needs of the garment making industry may well see the demise of the last of the conventional looms before long.

The layout of the whole of the finishing departments was also brought under review; time and money will certainly bring about a major reorientation in this area. In the weaving shed itself efficiency had been falling. Changes in beam sizes and greater variety could be expected to bring about a decline, but the loss was greater than had been expected, so MSC were commissioned to carry out a study of the shed and report. That document, when discussed by the Board in April 1987, was deemed inadequate, and sent back to be done again.

In the yarn sales field Stuart Allen replaced Allen Hughes as our agent in Eire, and Gardiner achieved an invitation to show at the Italian Yarn show, Inter Filati. The drive to be seen at shows, for yarn or cloth, was taking hold.

In May 1987 the last tungsten lighting in the mill was replaced by fluorescent equipment, 41 years after tungsten had replaced gas.

The year to 31 January 1988, the first full year since the old men had retired, showed a spinning profit of £307,000 and a weaving profit of £272,256, giving a total pre-tax result of £579,256 which, but for two errors of judgement towards the end of the year might well have been well over six hundred thousand. The dividend was set at 15p per share.

That year was soon to be seen as having incorporated the best of the prevailing textile cycle and the Board meetings in February and March devoted time to dealing with the first, almost inevitable consequence of the onset of a downturn, overstocking in both yarn and fibre. Strenuous action was demanded which, by May was bringing results in the form of reduced stock levels. By that meeting other signs of impending recession were showing, with a declining weaving order book forcing the reduction of the length of weaving shifts from ten hours to eight "as soon as the order book permitted". By June the sales order position, both cloth and yarn, was being looked at anxiously, a major yarn customer had deliberately delayed payment of £99,000 and the Scottish Industry was in nation wide dispute with the Trades Unions about the interpretation of agreements on short time working. The clouds of a downturn were gathering. The Chairman called for management to produce a survey of plans and prospects covering the next two years. That was ironic; as non executive directors we were calling on management to do what we had always refused to do for Tootal; make forecasts and budget profits!

In June 1988 Gigi Saada, our agent in Paris whose performance over the years had been erratic, to put it at its best, was given notice, and set about starting legal proceedings for substantial compensation. EEC rules for such compensation were at the revised draft stage and the legal position was a trifle uncertain. The claim was ultimately settled for £6,500 and kept out of the courts. Bernard Floux was appointed to succeed Saada.

In New York the linchpin of our operation through 18 International was a Yorkshireman by the name of Kevin Taylor, who had advised on styling and maintained customer contact between Peter Harvey's frequent visits. In June he left to go to work for a large Canadian firm, leaving Peter to watch the future handling of our account by 18th. with some trepidation. Troubles never come singly.

The half year to 23 July 1988, or rather the 25 weeks to that date, showed a spinning profit of £206,627 and a weaving profit, after charging against it the overall management commission of £42,811, of £156,855, a total of £363,482, somewhat less than the £471,000 achieved in the previous corresponding half, but still an excellent result.

The rest of this year,, and of this history, is devoted to the sale of Gardiner of Selkirk to S. Jerome and Sons Ltd. of Shipley. The sale of Gardiner had been an objective, more urgently sought by Tootal than by the private members, but sought by them just the same, so when in June of 1988 Alan Webb, Finance Director of Tootal, told us that Jeromes had approached Tootal with a view to purchase, we thought "here we go again", not knowing that this was the one which would come to fruition.

Once again Colin and I ventured down the M6 to Manchester to discuss with Alan Webb and his team the value of Gardiner and, after a mornings debate, calculation and crystal gazing came up with a figure which we regarded as both an opening shot and a sticking point. After all Jeromes had access to exactly the same P/E ratios and yields, printed daily in the Financial Times, as we did. Any such figure is the result of a large number of compromises involving asset values, profits past and forecast, the state of trade, the keenness of the members to get out and the divined anxiety of the purchaser to acquire. Whatever the origins of the figure it was handed to Richard Gilmore, Tootal's executive handling disposals of assets, to achieve if he could.

Sooner than we expected Richard reported that Jerome were willing to offer that figure and, the price agreed, the negotiations entered the formal reporting by accountants and turning over stones phase. Gardiner's 23 July accounts were rushed out and, contrary to our usual practice for half years accounts, audited. For Jerome, Grant Thornton analysed them and spent days in Selkirk, looking over every aspect of the Gardiner organisation. Valuers surveyed, reported on and valued the buildings and plant. The new FAX machine spewed reports on overdue debtors, stocks, creditors and anything else which might possibly hide a nasty. Richard Gilmore, Gordon Tregaskis, Tootal's Group Legal Adviser and Jonathan Shorrock, of Addleshaw, Sons and Latham, Tootal's Manchester Solicitors, all came up to Selkirk to go over and adjust the Warranties required by the purchaser, and to write the Disclosure Letter in which the vendors declare facts not covered by the Warranties. It all took a lot of time and unaccustomed effort, but in the end it was done. Breach of warranty involves penalty in the form of repayment of part of the purchase price and the last hurdle which we had to overcome was Jerome's insistence that Colin and I should be "Jointly and Severally" responsible for such payments in respect of claims related to our separate shareholdings, a demand which we simply could not accept. Fortunately Jerome did not regard it as a sticking point.

The contracts were signed on first October 1988. All three vendors were insistent that they be paid in cash, not shares, and this was achieved by the issue of Jerome shares and their placing with institutions immediately.

On 31 October 1988 I, with Colin Brown, Gordon Tregaskis and Jonathan Shorrock, went to Leeds to Singer and Freidlander's offices in Park Row to sign Gardiner of Selkirk Ltd. away. There was a Board meeting at which A. H. Jerome, R. H. Jerome and S. M. Jerome were appointed Directors, immediately followed by the resignations of H. C. Brown, I. McK. Jackson, F. M. Brown and J. M. Jackson. Transfers of the whole Gardiner capital to Jerome were noted, and the meeting closed.

All that remained to do was for the vendors to sign away to the institutions with whom they had been placed the 787,234 Jerome Ordinary shares which had been issued to them, whereupon the cash equivalent was banked in their bank accounts.

So ended a long, rewarding and profitable conduct of Gardiner's affairs in the names of Tootal, Brown and Jackson. Gardiner the firm goes on as a subsidiary of Jerome. Perhaps another thirty years on someone else will bring this tale up to date. And why not? The wool textile trade has never been easy, and often seems to be in a terminal state but yet goes on. It was ever thus. Colin Brown's great grandfather Henry Brown wrote in his journal in 1828: -

" Too thin does not please, too stout does not pay. Galashiels cloth has the appearance of going out of fashion".